Statutory Adjudication

A Practical Guide

Statutory Adjudication
A Practical Guide

Derek Simmonds
CEng, BSc, FICE, FCIArb, FCMI, FConsE
Registered and Chartered Arbitrator

Blackwell
Publishing

Blackwell Publishing Ltd
Editorial Offices:
9600 Garsington Road, Oxford OX4 2DQ
 Tel: +44 (0) 1865 776868
108 Cowley Road, Oxford OX4 1JF, UK
 Tel: +44 (0)1865 791100
Blackwell Publishing USA, 350 Main Street,
Malden, MA 02148-5018, USA
 Tel: +1 781 388 8250
Iowa State Press, a Blackwell Publishing
Company, 2121 State Avenue, Ames, Iowa 50014-
8300, USA
 Tel: +1 515 292 0140
Blackwell Munksgaard, 1 Rosenørns Allé, P.O.
Box 227, DK-1502 Copenhagen V, Denmark
 Tel: +45 77 33 33 33
Blackwell Publishing Asia Pty Ltd, 550 Swanston
Street, Carlton South, Victoria 3053, Australia
 Tel: +61 (0)3 9347 0300
Blackwell Verlag, Kurfürstendamm 57, 10707
Berlin, Germany
 Tel: +49 (0)30 32 79 060
Blackwell Publishing, 10 rue Casimir Delavigne,
75006 Paris, France
 Tel: +33 1 53 10 33 10

First published 2003

A catalogue record for this title is available from
the British Library

ISBN 1-4051-1085-6

Library of Congress
Cataloging-in-Publication Data
is available

Set in 10.5/12.5pt Palatino
by DP Photosetting, Aylesbury, Bucks
Printed and bound in Great Britain by
MPG Books Ltd, Bodmin, Cornwall

For further information on
Blackwell Publishing, visit our website:
www.blackwellpublishing.com

Contents

Preface   xiii
Abbreviations   xv

Part 1 An Introduction to Adjudication   1

1.   **Adjudication – a brief resumé**   3
1.1   What is adjudication?   3
1.2   Construction industry adjudication pre-1998   3

2.   **The need for change in the construction industry**   5
2.1   The investigation of Sir Michael Latham   5
2.2   The Housing Grants, Construction and
          Regeneration Act 1996   5
2.2.1   Statutory adjudication   6
2.2.2   Statutory payment requirements   6
2.3   The Scheme for Construction Contracts
          Regulations 1998   6
2.4   Contract adjudication provisions   7
2.5   Possible amendments to legislation   8

3.   **Some adjudication terminology**   9
3.1   Referral   9
3.2   The parties   9
3.3   Adjudicator nominating body   9
3.4   The adjudicator   10
3.5   Jurisdiction   11
3.6   The decision   11
3.7   An adjudication day   11

PART 2 So You Want To Go To Adjudication?   13

1.   **Is referral to adjudication really in your best
          interests?**   15
1.1   Have you exhausted negotiation?   15
1.2   What about mediation or conciliation?   15
1.3   Is adjudication suitable for your dispute?   16

Contents

1.4   Should you avoid adjudication and go directly to arbitration or litigation?   16

2.   Your entitlement to refer a dispute to adjudication under the Housing Grants, Construction and Regeneration Act 1996   17
2.1   When can you instigate adjudication?   18
2.2   How long will it take?   18
2.3   Do you have a contract?   19
2.4   Do you really have a referable dispute?   19
2.4.1   Establishing the existence of a dispute   20
2.4.2   Restriction on more indirect disputes   21

2.5   Does the Act apply to your contract?   21
2.5.1   When did you enter contract?   21
2.5.2   Do you have a contract in writing?   22
2.5.3   Is yours a construction contact?   24
2.5.4   What are construction operations?   25
2.5.5   What are not construction operations?   26
2.5.6   What if your contract is mixed?   27
2.5.7   Where is your contract being carried out?   27
2.5.8   Exclusions   27
2.5.8.1   Residential premises   28
2.5.8.2   Other   28

2.6   Other considerations   29
2.6.1   Does the law of the contract make any difference?   29
2.6.2   What if your dispute is already the subject of some other dispute resolution or legal process?   30
2.6.3   What if your contract has been allegedly repudiated or has been terminated?   30
2.6.4   Can you adjudicate on more than one dispute at the same time?   30
2.6.5   Duplication of referrals   31
2.6.6   Settlement   32
2.6.7   Can the responding party refuse or delay adjudication?   32
2.6.8   What if the responding party is in financial difficulties?   33

2.7   Summary   34

3.   What next?   34
3.1   Which adjudication procedure?   34
3.2   Do you need to engage a lawyer or consultant?   35

Contents

4.	**What will it cost?**	**36**
4.1	The adjudicator's costs	36
4.1.1	The adjudicator's hourly rate	37
4.1.2	The size and complexity of the dispute	37
4.1.3	The ability and experience of the adjudicator	37
4.2	The parties' costs	38
4.3	Special provisions as to costs	39
5.	**Getting the adjudication under way**	**40**
5.1	With whom or to where do you correspond?	40
5.2	The notice of adjudication	41
5.3	Getting an adjudicator in place	43
5.3.1	Named in the contract	43
5.3.2	Agreed between the parties	43
5.3.3	Nominated by an adjudicator nominating body	43
5.4	The adjudicator's Contract, Agreement or terms and conditions	48
5.5	What happens if there is a delay in the appointment of an adjudicator?	50
5.6	The referral notice	50
5.6.1	General considerations	50
5.6.2	Format of the notice	51
5.6.2.1	Introduction	52
5.6.2.2	Narrative	52
5.6.2.3	Redress	53
5.6.2.4	Supporting documentation	54
5.6.2.5	Summary	56
5.7	What if you are unable to submit the referral notice on time?	56
6.	**The conduct of the adjudication**	**57**
6.1	The process after referral	57
6.1.1	The responding party's response to the referral notice	59
6.1.2	Your reply to the response	59
6.1.3	Further submissions	60
6.2	Refusal of the responding party to participate	60
6.3	Can you withdraw the referral notice?	61
6.4	Can the responding party submit a counterclaim?	62
7.	**The decision**	**62**
7.1	Format of the decision	62
7.2	A reasoned decision	63
7.3	Obtaining the decision	64

7.4	What if the adjudicator has not dealt with all the issues referred to him?	65
7.5	What if you consider the adjudicator's costs to be excessive?	66
8.	**Enforcement of the decision**	**66**
8.1	What if the responding party does not do as the adjudicator has ordered?	66
8.2	What if the responding party contends that they are unable to pay you?	69
8.3	What if the responding party refuses to pay on the grounds of a set-off or counterclaim?	70
8.4	Would enforcement proceedings be successful if you became insolvent?	71
PART 3 So You Are Being Taken To Adjudication		**73**
1.	**Initial steps**	**75**
1.1	What do you do when you find you may be taken to adjudication?	75
1.2	Can you refuse to participate in an ajudication?	75
1.3	What do you do when you receive a notice of adjudication?	76
2.	**Objecting to becoming involved**	**77**
2.1	What can you do if you consider you have a valid objection to the adjudication proceeding?	77
2.1.1	Stand aside, let him and the referring party proceed without you and then refuse to carry out the decision if it goes against you	79
2.1.2	Participate without prejudice to your view that you are not required to do so, and then refuse to carry out the decision if it goes against you	79
2.1.3	Seek the intervention of the court	80
2.2	What if you have reservations about the nominated adjudicator?	80
2.2.1	Technical qualifications	80
2.2.2	Adjudication experience	81
2.2.3	Lack of impartiality	81
2.3	What if the referring party is in financial difficulties?	82
3.	**Active participation**	**83**
3.1	Should you be represented?	83
3.2	Agreeing an adjudicator	83

Contents

3.3   What if the matters in the referral notice are
        different from those in the notice of adjudication?   84
3.4   Your response – your defence to the referral notice   85
3.5   Can you request a meeting?   87
3.6   Can you submit a counterclaim?   88
3.7   Can you settle during the course of the adjudication? 89
3.8   What if the referring party asks the adjudicator for
        recovery of its costs?   89

4.   When the decision is ready   90
4.1   When do you get to see the decision?   90
4.2   Is it necessary to contribute to the adjudicator's
        charges?   90
4.3   What if the referring party does not take up the
        decision?   90
4.4   What if you are unable to pay what the adjudicator
        has ordered?   91

5.   Challenging the decision or its enforcement   91
5.1   Resisting because you consider the adjudicator
        lacked jurisdiction   92
5.2   Resisting because you intend to refer the dispute
        to arbitration or litigation   92
5.3   Resisting because the referring party is or will be
        indebted to you and is in financial difficulties   92
5.4   Possible actions to resist complying with the
        adjudicator's decision   94

PART 4 Matters of Common Interest   95

1.   During the currency of the adjudication   97
1.1   Resourcing the adjudication   97
1.2   Concerns regarding the competence of the
        appointed adjudicator   98
1.3   Concerns regarding the behaviour of the adjudicator   99
1.4   Can you be required to give evidence before
        the adjudicator?   101
1.5   Should there be a site inspection   102
1.6   Are you required to comply with the demands of
        the other party?   102
1.7   Can you object to the directions of the adjudicator?   103
1.8   What if you are unable to comply with the
        directions of the adjudicator?   103
1.9   What if new material is introduced?   104

Contents

1.10	If you get out of your depth can the adjudicator assist?	105
1.11	Can you seek the intervention of the courts?	106
1.12	Can the adjudication be suspended?	106
1.13	What scope does the adjudicator's expert have?	106
1.14	What if the adjudicator is late in giving his decision?	107
1.15	Requesting reasons	108
1.16	What if the adjudicator fails to make a decision?	108
2.	**After the adjudicator has released his decision**	**108**
2.1	What happens if you do not like what you see in the decision?	108
2.2	Can the adjudicator be required to explain how he reached his decision?	109
2.3	Slips and errors in the decision	110
2.4	What if you consider the adjudicator's costs to be excessive?	112
2.5	Joint and several responsibility for the adjudicator's costs	113
3.	**Challenging the decision**	**113**
3.1	If you conclude that the adjudicator has not acted impartially	114
3.2	If you consider that the adjudicator has exceeded his jurisdiction	114
3.3	If you consider the adjudicator has seriously misunderstood the dispute	115
3.4	The process leading to a challenge of the decision or its enforcement	116
4.	**The final solution**	**116**
4.1	Settlement	116
4.2	Arbitration or litigation	116
PART 5	**Statutory Payment Requirements**	**121**
1.	**Resumé of the requirements of the HGCRA and related regulations of the Scheme for Construction Contracts**	**122**
1.1	Act: section 109; Scheme: paragraphs 1, 2	122
1.2	Act: section 110; Scheme: paragraphs 3–9	122
1.3	Act: section 111; Scheme: Paragraph 10	124
1.4	Act: section 112; Scheme	125
1.5	Act: section 113; Scheme: paragraph 11	126

Contents

**PART 6 Comparison of the Scheme for Construction
Contracts and other principal adjudication
provisions and procedures**   **129**

1.   The Scheme for Construction Contracts
(England and Wales) Regulations 1998   132

1.1   The Scheme for Construction Contracts in Northern
Ireland Regulations (Northern Ireland) 1999   136

1.2   The Scheme for Construction Contracts (Scotland)
Regulations 1998   136

2.   The ICE Conditions of Contract, 7th Edition and
Adjudication Procedure (1997)   136

3.   The JCT Standard Form of Building Contract 1998,
Amendment 4, 2002   141

4.   The Engineering and Construction Contract,
Second Edition and Option Y(UK)2 1998   143

5.   Institution of Chemical Engineers Form of Contract,
Third Edition 2002 and Adjudication Rules,
Second Edition 2001   147

6.   GC/Works/1 (1998) – General Conditions   149

7.   The Technology and Construction Solicitors
Association 2002 Version 2.0 Procedural Rules
for Adjudication   151

8.   The Construction Industry Council Model
Adjudication Procedure, Second Edition   155

9.   Subcontract forms   157

Appendices   **159**

1.   The Housing Grants, Construction and
Regeneration Act 1996   159

2.   The Construction Contracts (England and Wales)
Exclusion Order 1998   171

3.   The Scheme for Construction Contracts
(England and Wales) Regulations 1998   177

4.   Contract Adjudication Provisions and Related
Procedures   191

4.1   ICE Conditions of Contract, 7th Edition (Extract)
ICE Adjudication Procedure (1997)   193

4.2   JCT Standard Form of Building Contract 1998,
Amendment 4, 2002 (Extract)   205

4.3   The Engineering and Construction Contract, Second
Edition (Extract) Option Y(UK)2 1998 (Extract)   213

5.   Independent Adjudication Rules and Provisions   221

Contents

5.1   Technology and Construction Solicitors Association
2002 Version 2.0 Procedural Rules for Adjudication   223
5.2   Construction Industry Council Model Adjudication
Procedure, Second Edition   233
6.   Adjudicator Nominating Bodies   245
7.   Alphabetical List of Cases   247
8.   Categorised List of Cases   255

Index   267

Preface

Unquestionably, statutory adjudication has made a significant impact on the construction industry since its introduction on 1 May 1998.

Although fewer than 50 nominations were made by the principal adjudicator nominating bodies throughout the remainder of 1998, nominations rapidly increased thereafter and by the end of 2002 the total number was around 6000. The number of adjudicators agreed by the parties is not known.

Original concerns regarding enforcement proved to be unfounded when the first challenge to an adjudicator's decision reached the court in February 1999. The decision was enforced swiftly through the summary judgment process and that set the pattern for all enforcement proceedings. The courts have responded to the spirit of the Act.

From time to time various attempts have been made to frustrate the adjudication process but most have failed. The courts have tended to look for ways of upholding a decision rather than finding reasons for setting it aside, albeit some rough justice has been dispensed along the way.

When adjudication first became the subject of debate and discussion – even prior to the passing of the Act in 1996 – it was contemplated that disputes referred would comprise relatively uncomplicated matters – a workmanship problem, an extension of time, the interpretation of a contractual provision and so on. Potential adjudicators were advised that they could effectively refuse complex disputes by deciding that they were not appropriate for the adjudication process and were therefore 'unadjudicable'.

In the event adjudication grasped the nettle of large complex disputes; indeed many adjudicators have found themselves deciding detailed final accounts, albeit taking full advantage of the provisions for extending the time allowed for them to do so.

Apart from the inevitable proliferation of seminars, courses and comment, statutory adjudication has generated about a dozen books. Generally these are directed either to professionals representing parties or to adjudicators themselves.

In contrast, this book has been written for the parties to adjudication. Although adjudication was built on the concept that the parties could represent themselves, parties may wish to or feel they should be represented. It is therefore important that they have an understanding of what is involved and how the process is conducted.

Part 1 outlines the background and development of statutory adjudication. Part 2 leads a party contemplating the referral of a dispute through the stages of the adjudication process, including the presentation of a referral submission, to enforcement. Part 3 takes a party faced with adjudication through the process, drawing attention to the various actions necessary to protect its interests, and on to the challenging of the decision. In the interests of avoiding or minimising duplication, Part 4 addresses matters which are of common interest.

The bedfellows of statutory adjudication are the statutory payment regulations included in the Housing Grants, Construction and Regeneration Act and designed with the object of eradicating abuse of what were often already unfair contractual payment provisions. These and their translation into the Scheme for Construction Contracts are dealt with at Part 5. The most frequently used adjudication provisions and procedures are compared at Part 6.

This book would not have been possible but for the competent aid of my secretary. Her patience and skills have been invaluable to me. Additionally I am grateful to Julia Burden, Deputy Divisional Director of the publishers, for her advice and guidance. I am also grateful to the following for their kind permission in allowing me to reproduce the documents appearing in the Appendices: the Royal Institute of British Architects (Appendix 4.2), Thomas Telford (Appendices 4.1 and 4.3), the Construction Industry Council (Appendix 5.2) and the Technology and Construction Solicitors Association (Appendix 5.1).

Where the male pronoun is used to refer to an adjudicator, this should be taken as representative of both male and female.

31 March 2003

Abbreviations

ACA	Association of Consultant Architects
ACE	Association of Consulting Engineers
ANB	Adjudicator nominating body
CC	Construction Confederation
CIArb	Chartered Institute of Arbitrators
CIC	Construction Industry Council
ECC	Engineering Construction Contract
HGCRA, The Act	Housing Grants, Construction and Regeneration Act 1996
ICE	Institution of Civil Engineers
IChemE	Institution of Chemical Engineers
JCT	Joint Contracts Tribunal
NEC	New Engineering Contract
NNB	Named nominating body
NSCC	National Specialist Contractors Council
ORSA	Official Referees Solicitors Association
RIBA	Royal Institute of British Architects
RICS	Royal Institution of Chartered Surveyors
TeCSA	Technology and Construction Solicitors Association
The Scheme	Scheme for Construction Contracts (England & Wales) Regulations 1998

Part 1
An Introduction to Adjudication

PART 1
AN INTRODUCTION TO ADJUDICATION

1. Adjudication – a brief resumé

1.1 What is adjudication?

Adjudication is a process whereby, when a dispute between con-tracting parties arises, a neutral – that is, basically, a person who has no connection with either side – is engaged to examine the argu-ments of the parties and to decide the dispute.

The decision may be temporary pending final determination by some other process or it may become final and binding if the dispute is not referred on within a specified period or if the parties have agreed that it should. Under certain types of contract, adjudication is a mandatory pre-step before the final process may be commenced.

1.2 Construction industry adjudication pre-1998

Adjudication in one form or another has been around for some time. For example in 1997 it was introduced into various forms of building subcontract such as DOM/1 and NSC4 (now NSC/C). The Association of Consultant Architects and the British Property Fed-eration Forms of Building Agreement, produced in 1982 and 1984 respectively, both included adjudication as an option available for a particular contract. The New Engineering Contract (now the Engi-neering Construction Contract) first published in 1993 followed a practice common under World Bank funded international contracts, of making adjudication a mandatory pre-step to arbitration.

Whereas the early adjudication provisions in the building sub-contract forms were designed solely to combat unwarranted set-off by the contractor (or at least to prevent the contractor benefiting from setting off improperly), the intention in other contract conditions was to try to provide a quick, inexpensive solution to any dispute which, hopefully, the parties would accept. If either party could not accept the adjudicator's decision, at least the issues would have been clarified and defined prior to any subsequent legal process. To encourage finality some contractual provisions imposed a time limit within which arbitration or litigation had to be commenced otherwise the adjudicator's decision would become final.

There appear to be no useful statistics as to the success of these earlier adjudication provisions. Relative to the huge number of building subcontracts let between 1977 and the introduction of statutory adjudication in 1998, only a handful of set-off disputes were referred to adjudication. No doubt the drafters of the provisions would like to think that the availability of a quick adjudication decision acted as a deterrent to unscrupulous contractors who might otherwise have set off unjustifiably. The truth is probably that contractors tended to avoid DOM/1 as its provisions did not accord with the way in which they preferred to operate their subcontracts; and the changing trends in the building industry led to a marked reduction in the use of nominated subcontracts and hence of form NSC/C.

The Association of Consultant Architects and British Property Federation forms were themselves infrequently used so utilisation of their adjudication provisions would have been minimal.

Under the New Engineering/Engineering Construction Contracts, adjudication was, and indeed still is, a mandatory first step prior to arbitration. The fact that very few arbitrations arose out of these forms suggests that most disputes did not continue beyond the adjudication stage, but this may be due more to the success of the form of contract in preventing disputes arising rather than to adjudication resolving them.

The drawback with adjudication in its various earlier forms was always that of enforcement of the adjudicator's decision. Adjudication being a contractual provision, failure of the losing party to comply with a decision is a breach of contract. In contrast to arbitration awards the process of obtaining a summary judgment in the courts was not, or appeared not to be, available for the enforcement of adjudicators' decisions.

However, regardless of their success or otherwise, all existing contractual adjudication provisions had to be abandoned once statutory adjudication was introduced since none of them complied with the minimum requirements laid down by the new legislation.

2. The need for change in the construction industry

2.1 The investigation of Sir Michael Latham

Various investigations into the construction industry over the years have concluded amongst other things that a disproportionate amount of time and money was being expended by contracting parties in resolving disputes between them.

Over all the construction industry none of the alternative dispute resolution procedures – mediation, conciliation or adjudication (as it then existed) – were having any noticeable effect on reducing the time and cost expended on disputes. The improper withholding of monies due was causing severe cash flow problems in the industry. This existed throughout the contractual chain starting with employers and running through to sub-subcontractors. DOM/1 and NSC/4 had attempted to address the problem only as it affected subcontractors.

An investigation in the early 1990s by Sir Michael Latham culminated in his report of 1994, *Constructing the Team*, in which he recommended that some quick simple procedure to provide at least a temporary resolution of a dispute should be available to the parties to construction contracts, and that there should be some means of preventing indiscriminate withholding by one party of monies due to the other.

The result was legislation in the form of Part II of the Housing Grants, Construction and Regeneration Act 1996 (other Parts cover unrelated matters) which addressed both of the principal Latham recommendations. The full title being something of a mouthful, it is referred to as the HGCRA or just the Act and it applies to England, Wales and Scotland. Northern Ireland has its own similar legislation, the Construction Contracts (Northern Ireland) Order 1997.

2.2 The Housing Grants, Construction and Regeneration Act 1996

The Act not only applies to disputes between clients and contractors or contractors and subcontractors, but also embraces service contracts between clients and their designers, quantity surveyors or other advisers.

However, not all construction contracts are caught by the Act; certain types of work are excluded as are contracts for supply only, but supply and fix contracts are caught by the Act.

Part II of the Act is itself divided:

Sections 104–107 and 117	determine what sorts of construction contracts fall to be dealt with by the Act
Sections 114–116	cover matters common to both adjudication and payment
Section 108	deals specifically with adjudication
Sections 109–113	relate to payment matters

2.2.1   Statutory adjudication

The Act enables either party to a contract to require that a dispute be immediately referred to an adjudicator whose decision is binding until the dispute is finally determined by some arbitral or legal process or by the agreement of the parties. With certain exceptions, the Act applies to all construction contracts and the parties cannot contract out of it. Although passed in 1996, the Act did not come into force until 1 May 1998. Northern Ireland legislation became operative on 1 June 1999.

The Act lays down certain procedural and other requirements relating to the conduct of adjudication, and all construction contracts caught by the Act must contain at least these requirements.

2.2.2   Statutory payment requirements

The Act and its counterparts in Scotland and Northern Ireland go beyond requiring construction contracts to provide for adjudication. They also set down regulations to the effect that contracts must contain a clear structure in relation to payment together with procedures to be adopted before a paying party can withhold monies. Collectively these are commonly known as the payment provisions.

2.3   The Scheme for Construction Contracts Regulations 1998

If particular contractual adjudication provisions do not fully comply with the minimum requirements of the Act, provisions known as the Scheme for Construction Contracts Regulations 1998, commonly known as the Government Scheme or simply the Scheme, apply. It is not permissible to pick clauses out of the Scheme to make good any deficiencies in the adjudication provisions in contracts; these provisions have to be abandoned totally and the Scheme substituted.

The same is not the case as regards payment provisions. In the event of non-compliance it is possible to utilise the payment provisions of the Scheme piecemeal to make good any deficiencies in the contractual provisions.

Produced in 1998, there are three versions of the Scheme, for England and Wales, for Scotland and for Northern Ireland. All are essentially the same but the Scheme for Scotland is slightly different in content as it has to accommodate the different law north of the border in respect of enforcement procedure.

2.4   Contract adjudication provisions

Following the passing of the Act, the principal drafting bodies of standard forms of contract (the Joint Contracts Tribunal, Institution of Civil Engineers (ICE), etc.) immediately set to and produced adjudication provisions tailored to permit incorporation into their standard forms. This is done in one of two ways.

For example, the Joint Contracts Tribunal family of contracts and GC/Works/1 forms of contract incorporate adjudication provisions totally within the contract conditions. The ICE family of contracts incorporate by reference a separate procedure, the ICE Adjudication Procedure. New Engineering Contract documents utilise the Scheme for Construction Contracts.

The Conditions of Engagement of the Royal Institute of British Architects (RIBA), the Royal Institution of Chartered Surveyors (RICS) and the Association of Consulting Engineers (ACE) all incorporate by reference the Construction Industry Council (CIC) Model Adjudication Procedure. The associated Forms of Agreement in respect of subconsultant, planning supervisor and project manager prepared by the RIBA also use the CIC procedure.

The Technology and Construction Solicitors Association (TeCSA) has produced an independent procedure favoured by lawyers who draft ad hoc construction contracts.

All of these procedures, of necessity, go beyond the minimum requirements of the Act and each has its own peculiarities, but in general terms they follow the format of the Scheme.

Many organisations and large companies have produced bespoke adjudication procedures designed to be compatible with their own forms of contract. Some of these contain provisions which favour the producer and are unfair to the other party, but nevertheless they have been upheld by the courts.

Payment clauses of all existing forms of contract of whatever nature have also had to be amended in order to comply with the Act. Standard forms are now inevitably suitably drafted, but purpose-made and company own contracts may still not comply, in which event the Scheme comes into play to the extent that is necessary to ensure compliance.

Initially loose-leaf amendments to all the standard forms were published to ensure immediate compliance with the Act as regards both adjudication and payment; these were commonly known as the 1998 amendments. Over the intervening period most have been fully integrated by way of new editions of the standard forms.

In the interests of uniformity there is currently a move to encourage the use of the Scheme rather than bespoke adjudication provisions. The extent to which these will be abandoned in favour of the Scheme remains to be seen.

2.5   Possible amendments to legislation

It was always understood that a review of the Act and Scheme would be undertaken once sufficient experience of their operation was available.

The Act is primary legislation and can only be amended by another Act of Parliament. As was the case with statutory adjudication, it is common practice to combine various pieces of legislation into one Act but there is always difficulty in finding a suitable legislative vehicle and allocating parliamentary time for debate. Consequently, at the time of the review implemented by the Department of Trade and Industry (which took over responsibility for statutory adjudication matters from the Department for the Environment, Transport and the Regions) it was thought that it would not be possible to amend the Act; but the Scheme for Construction Contracts could be readily amended as that was secondary legislation not requiring an Act of Parliament.

Consultation took place with those involved with adjudication, and proposals for amending the process were considered. These were narrowed down and in August 2001 a draft amendment to the Scheme was prepared. This dealt with three issues:

- Parties' costs
- Correction of slips
- Requests for reasons.

Following further consideration it was decided that there would be only one amendment – that relating to parties' costs – but, whatever the difficulty, the Act would also be amended. It was recognised that only a relatively small proportion of adjudications were by that time being conducted under the Scheme (and encouragement to abandon other procedures in favour of the Scheme was not being measurably successful). Consequently there was little point in making any amendment to the Scheme unless the Act was also amended.

The latest information, at March 2003, is that a suitable legislative vehicle is still awaited in order to amend the Act; the Scheme will be amended as appropriate at the same time to reflect the new provision in the Act. The 2001 draft amendment relating to parties' costs was to have amended regulation 20 of the Scheme and would have prevented an adjudicator from taking into account any matter relating to their legal or other costs. The precise wording of the eventual amendment has not been settled but it is intended that it will prevent abuse of the process whereby the allocation of parties' costs in favour of the responding party may be predetermined by terms in the contract regardless of the outcome of the adjudication.

3. *Some adjudication terminology*

3.1 Referral

The process of having a dispute decided by an adjudicator is known as referral to adjudication or referral to an adjudicator.

3.2 The parties

The party wishing to commence adjudication proceedings is known as the referring party.

There is no universally accepted term for the other party but responding, non-referring or receiving party are commonly used.

Use of the terms claimant and respondent is discouraged as these relate to the parties at arbitration.

3.3 Adjudicator nominating body

The parties are free to agree anyone they wish to act as adjudicator but if they cannot agree then the referring party may approach what is known as an adjudicator nominating body (ANB).

At the time that statutory adjudication was being debated, the need to have some system in place should the parties be unable to agree an adjudicator was recognised. Independent bodies would have to be set up and it was intended that these would be subject to some sort of registration. An organisation wishing to become a nominating body would have to demonstrate that it could provide a service capable of meeting the requirements of the Act and that all those whom it might select would be suitable persons.

Registration of ANBs was abandoned before the Act came into force and, although the Construction Industry Council is able to exert some influence over the primary professional bodies within its ambit who make nominations, there are no statutory controls governing ANBs.

The Act makes no reference to the process of getting an adjudicator in place and the only restrictions are laid down by the Scheme, paragraph 2(3) of which effectively states that any body, not being a natural person nor party to the dispute, which holds itself out publicly as a body which will nominate an adjudicator if requested, may call itself an ANB. The Scheme uses the term select, but nominate is the expression most generally used. The term appoint is frequently used and also appears in the Scheme but there its use is confined to situations after the adjudicator is in place, for example the revocation of the appointment of an adjudicator.

The majority of standard forms of contract specify the ANB to be used in the absence of agreement. Most frequently this will be either the ICE, RIBA, RICS or Chartered Institute of Arbitrators (CIArb). All these bodies nominate only those who have undertaken a recognised training course and interview and require their adjudicators to undertake regular further training. The same is true of certain other ANBs but not all.

3.4 The adjudicator

The Act itself does not say so but an adjudicator has to be a natural person agreed by the parties or nominated by an adjudicator nominating body. The position is a personal one – a firm, group, partnership, etc. cannot be named or appointed as adjudicator. Even though the adjudicator is employed by, for example, a firm of architects, the firm does not become involved and has no authority over or responsibility for the adjudicator in respect of his adjudication activity. An adjudicator is not required to be qualified in any way but it would clearly be foolish to agree someone who was not technically and legally competent and who had not been through an approved course of training.

3.5 Jurisdiction

Jurisdiction is the authority of the adjudicator to decide the dispute put before him. An adjudicator cannot decide in the legal sense whether or not he has jurisdiction – that is something only the courts can do unless the adjudication procedure specifically gives him the authority or the parties subsequently empower him by agreement. If the point arises, the adjudicator will consider the relevant facts and make up his mind whether or not to proceed. If he does proceed the objecting party may seek an injunction to stop the adjudication or, more usually, his decision may be subsequently challenged in the courts on the ground of lack of jurisdiction.

3.6 The decision

The decision is the document produced by the adjudicator after he has completed his deliberations on the matters referred to him. It will inform the parties as to the extent, if any, that the referring party has been successful; or, if the interpretation of a contractual situation is being sought, then it will provide answers to the questions posed.

3.7 An adjudication day

The Act and all adjudication procedures require various things to be done within so many days. This means calendar days, i.e. Saturdays and Sundays are included. Thus nine days could mean only five normal working days if the period is spread over two week-ends.

However, the Act at section 116(3) states that Christmas Day, Good Friday and bank holidays are excluded.

Part 2
So You Want To Go To Adjudication?

PART 2
SO YOU WANT TO GO TO ADJUDICATION?

1. *Is referral to adjudication really in your best interests?*

Whatever it does achieve, adjudication is not a panacea for all the problems in the construction industry. Referral of a dispute to adjudication is not necessarily the best move and before doing so you should consider the alternatives.

1.1 Have you exhausted negotiation?

Negotiation of a dispute is the best way of resolving a problem whilst maintaining good relationships. It requires skill and patience; if you can persevere you will often achieve the best result. However, prolonged negotiation may become too time consuming – you find that you continually go over the same ground making little or no headway. And if cash flow is a problem you cannot accommodate protracted negotiations. It may be better to move on, otherwise you could find yourself making too many concessions in an effort to reach a settlement.

1.2 What about mediation or conciliation?

If negotiation has failed but you are concerned that good relationships with the other party are maintained, then you should probably try mediation or conciliation. There is an increasing adoption of these processes, particularly mediation, but they require the consent of both parties; mediation or conciliation will not succeed unless both sides are willing at least to try. By their very nature, particularly mediation, they move towards a compromise. So if you feel strongly that you have a good case, these processes are probably not for you.

1.3 Is adjudication suitable for your dispute?

Contrary to expectation at the time that statutory adjudication was introduced, the sort of disputes that have been referred to adjudication have often turned out to be extremely complex. Many adjudicators have been asked to decide final accounts containing numerous issues in dispute. This is not the sort of thing that adjudication was designed to address. The limited time-scale, even if extended, can prevent an adjudicator giving proper attention to a multitude of separate issues with the result that he will probably be unable to do justice to the task presented to him. The limits of adjudication in this respect are recognised – not least by the courts when they have been asked to look into adjudicators' decisions. The process is acknowledged as often being rough and ready. The adjudicator will produce a decision but how similar it would be to the award of an arbitrator with all the time and facilities that he has at his disposal is questionable. But even if you are successful to only a modest degree, you at least obtain something quickly and you can pursue your considered full entitlement later.

1.4 Should you avoid adjudication and go directly to arbitration or litigation?

Referring a dispute to adjudication is not good for relationships particularly if your case is based on a technical point, for example if the other party had forgotten to issue a withholding notice. Whilst you will have all the time you need to prepare your case, the other party must respond in what is inevitably a very short period. This throws a great strain on the other party's organisation (even if outside assistance is obtained) and will disrupt its work programme. The other party may well attempt to frustrate the adjudication by challenging the validity of the adjudication or the jurisdiction of the adjudicator. You will need to counter promptly the arguments put forward; once the adjudication is under way you are under as much pressure as the responding party. And in the end you may have to instigate enforcement proceedings to secure whatever the adjudicator orders.

You must therefore look at your case realistically and decide whether the effort and cost (which in most instances will not be recoverable) of adjudication will be worthwhile for the sake of what may turn out to be only a temporary success.

2. *Your entitlement to refer a dispute to adjudication under the Housing Grants, Construction and Regeneration Act 1996*

As a party to a construction contract your entitlement to refer a dispute to the decision of an adjudicator is enshrined in Part II of the Housing Grants, Construction and Regeneration Act 1996. The *dispute* is stated to include *any difference.*

section 108(1)

All contracts caught by the Act are required to contain certain minimum provisions, as follows.

section 108(2)–(4)

Time periods are set down for the appointment of an adjudicator and completion of the adjudication.

section 108 (2)(a)–(d)

The time-scale of an adjudication is discussed later in this Part, at 2.2

The adjudicator has a duty to act impartially.

section 108(2)(e)

Problems arising from the failure of adjudicators to be impartial are discussed at Part 4, sections 1.3 and 3.1, in this book.

In recent years arbitrators and judges have been encouraged to take a more active role in the process under their control. Because of the nature of adjudication, not least the short time-scale, it is particularly important that adjudicators are also prepared to be proactive. To assist in this they may, if necessary, ascertain the facts and the law.

section 108(2)(f)

The decision of the adjudicator is binding – that means the responding party has a legal obligation to comply with it. However, unless the parties agree to go further and to accept a decision as finally determining the dispute, either may refer the dispute to arbitration or litigation, in which case the award of the arbitrator or the court supercedes the adjudicator's decision.

section 108(3)

The adjudicator is protected from legal action by disgruntled losers. He is not liable for anything done or omitted in the course of his appointment unless as a result of bad faith on his part. Any employees or agent of the adjudicator are similarly protected.

section 108(4)

Should a contract not contain the minimum requirements of section 108(1)–(4), the adjudication provisions of the Scheme for Construction Contracts apply.

section 108(5)

Other provisions defining the framework of statutory adjudication are set down at sections 104–107 and 114–117 and in the Construction Contracts Exclusion Order 1998. These are discussed later in this Part, at section 2.5.

2.1 When can you instigate adjudication?

The Act states that you may commence adjudication proceedings at any time; subject to the statutory limitation period, that means as soon after or as long after a dispute has arisen as you wish – it is entirely up to you.

section 108(2)(a)

As the referring party you have a great advantage in that you can take as long as you need to prepare your submission before you give a notice of adjudication.

Of course if you have settled everything and have been paid the amount of the final account or equivalent, that will normally be evidence of the resolution of any disputes you may have had and there should be nothing left on which to adjudicate.

You cannot open up any lawful and binding agreement you may have made with the other party. But the fact that your contract has been terminated makes no difference. Nor does it matter that the dispute is already the subject of arbitration or litigation proceedings, conciliation or mediation; you can still refer it to adjudication without affecting the other process and without the adjudication being affected by the other process[79]. (See later in this Part, at sections 2.6.2 and 2.6.6.)

2.2 How long will it take?

The objective of the Act is that the period from when you advise the other party of your intention to refer a dispute to adjudication (the notice of adjudication), to getting an adjudicator in place and stating to him what you want him to decide, should be no more than seven calendar days.

section 108(2)(b)

18

This is normally possible provided you do what you have to promptly and correctly. However, it is clear that it may not be achievable if difficulties arise.

The adjudicator has 28 days from the time the dispute is referred to him in which to produce his decision.

If for whatever reason he needs more time you, as the referring party, can give him a further 14 days.

section 108 (2)(d)

If serious difficulty arises and the adjudicator asks for yet more time, the consent of both parties is required.

section 108(2)(c)

Although the majority of adjudication decisions are produced within the 28 days, quite a few take longer on account of the increasing tendency for complicated disputes to be referred; two to three months is not uncommon. At the other end of the scale, however simple the dispute, it is rarely that an adjudicator produces his decision in under 21 days.

2.3 Do you have a contract?

You must have a contract with the party with whom you are in dispute. It may seem strange but the existence of a contract is not always certain and such uncertainty itself may be the subject of dispute[83/127]. When a company or practice has amalgamated, been taken over or simply changed its name, there can be confusion as to who exactly are the parties to the dispute[46].

Whilst you can request an adjudicator to decide what are the terms of your contract, he cannot determine whether or not you actually have a contract[21]. That amounts to asking him to decide whether he has jurisdiction and he is not allowed to do that unless the operative adjudication provisions specifically empower him or the parties have agreed that he may do so[77/116/125].

2.4 Do you really have a referable dispute?

On one of the rare occasions that a court has been asked to intervene in adjudication proceedings, a judicial review concluded there was a referable dispute as the parties had had some months to consider their respective positions[20]. In contrast a number of parties have been unsuccessful at enforcement proceedings because the courts have held that the adjudicator was in error and at the time of referral there was no dispute[101].

2.4.1   *Establishing the existence of a dispute*

For a dispute to exist, a party must have made a claim, request, complaint or allegation of some kind which has not been addressed or which has been rejected by the second party. The party intending to refer the dispute must have made it reasonably clear to the other what it is aggrieved about and to have provided at least some indication as to what redress it is seeking. The responding party is entitled to know with some certainty what is being sought from it.

Misunderstanding as to whether a dispute exists arises principally in connection with payment matters. For example, if you make an application for payment and it is not paid or monies are withheld – you consider improperly – then clearly you have a dispute. Indeed it is not necessary to wait until the due date for payment before giving notice of adjudication if the other party has clearly indicated by its actions or lack of action that it does not intend to pay your considered entitlement[34]. But if you are a contractor and have merely stated in general terms that, for example, you are incurring extra costs or will be seeking re-rating of certain activities, or you are an architect and you simply warn your client that you will be incurring additional fees for extra design work, then you have not as yet made a claim which the other party is able to address. You have never properly claimed the monies you wish to recover, no dispute can have arisen and hence there is nothing to refer.

Similarly, for example, if you are a subcontractor and have made vague references to a considered entitlement to an extension of time, perhaps as a defence to a criticism of lack of progress of your work, a dispute will not have been established simply because an extension has not been given. You must ensure that any process specified in the subcontract has been complied with and your request refused (in whole or in part) or ignored, before a dispute can be said to have arisen.

If a claim of some sort has been registered by you (or someone on your behalf[34]) then, subject to the terms of the contract, the responding party is entitled to a reasonable time in which to respond[101]. If a reasonable request for further information is made, a dispute will not arise until that information is considered and rejected[126]. However, unreasonable delay in the provision of an appropriate response has been held to give rise to a dispute[42].

The trend of judgments where the existence of a dispute has been in question is towards a more strict interpretation. The courts need to be satisfied that a dispute has crystallised[42].

2.4.2   *Restriction on more indirect disputes*

If the Scheme applies, your dispute must have arisen *under* the contract (paragraph 1). Other procedures broaden the range for referable disputes by use of terms such as *out of*, *relating to* or *in connection with*.

2.5   Does the Act apply to your contract?

Your contract may contain provisions for adjudication but for some reason will not be caught by the Act. Perhaps this is because the person preparing the tender documents has failed to delete these provisions. In that event there is nothing to prevent you from pursuing your dispute through adjudication under the contractual provisions. Indeed, certain forms of contract which provided for adjudication prior to 1998, for example the Engineering Construction Contract (ECC), may require you to do so before you can proceed to arbitration. The adjudication will be conducted in accordance with the contractual provisions and, unless the form of contract predates the commencement of statutory adjudication, 1 May 1998, these will probably reflect the requirements of the Act.

Since the introduction of statutory adjudication the courts, when asked, have tended to look upon non-statutory adjudication in much the same way[69/86/90]. But, as there are always exceptions[129], if you have a dispute that you have been unable to resolve, it is important to establish whether your contract falls under the Act.

This in itself may be a tricky exercise.

2.5.1   *When did you enter contract?*

The Act does not apply to contracts entered into prior to 1 May 1998.

It is normally a straightforward matter to decide the date of contract but complications can arise if there has been a letter of intent, an oral agreement to proceed or if there is no documentary evidence, etc. As to when a legal contract may have been formed, this will depend on the facts.

There have been a number of cases where the courts have considered whether the date of contract preceded 1 May 1998. Even though many terms had been agreed and work had actually started, where a contract had not been concluded until after 1 May it was held that it was caught by the Act[119]. In another case the parties had signed a form of contract before 1 May 1998 but, on the facts, the court held that there was a lack of contractual intention to enter into contract at that time, a contract not being entered into until a year later[6].

Where a letter of intent before 1 May had been superseded by a contract after 1 May, the Act was held to apply[28]. And entering into a novation agreement after 1 May 1998 may render disputes arising out of a contract before 1 May eligible for referral to adjudication[120]; but a deed of variations entered into after 1 May was not sufficient to bring an earlier contract within the scope of the Act[37].

Over five years on it is unlikely that the date of contract will still be an issue. However, yours may be a long running contract and if you are in doubt you should obtain legal advice, but ensure that your adviser is familiar with construction contract matters. A small provincial solicitor may be excellent for dealing with your routine company or partnership affairs but may be totally out of his depth with regard to contractual problems of this nature.

2.5.2 *Do you have a contract in writing?*

To be caught by the Act your contract must be in writing.

section 107(1)

Many contracts particularly smaller ones are not the subject of an actual agreement in writing, but the Act construes *agreement in writing* very broadly. In fact it effectively adopts the interpretation in the Arbitration Act 1996. The result is that it is virtually impossible for any contract of substance to escape the Act on the grounds that it is not in writing, but there have been some exceptions.

There is an agreement in writing in the following eventualities

section 107(2)

(a) *if the agreement is made in writing (whether or not it is signed by the parties)*
    A situation frequently arises whereby a contract document or agreement is prepared but not sent to the other party or not returned by them. The reasons for this are varied but, provided there is a valid contract, the lack of signature is of no consequence when determining whether it is subject to the requirements of the Act.

(b) *if the agreement is made by an exchange of communications in writing*

A common arrangement for smaller contracts is for offer and acceptance by letter.

(c) *if the agreement is evidenced in writing*
This would accommodate, amongst other things, a situation where the agreement has been lost but there is some other document which confirms that it was indeed made.

If an agreement is not actually made in writing it is, for the purpose of interpretation of the Act, made in writing if terms which are in writing are referred to.

section 107(3)

Evidence in writing may be by one of the parties or any authorised third party recording the agreement.

section 107(4)

An agreement in writing is deemed to arise if, in an exchange of written submissions in adjudication, arbitral or legal proceedings, one party alleges the existence of an agreement otherwise than in writing and that allegation is not denied by the other party in its response.

Reference to *in adjudication* applies, it may be thought, to the ongoing adjudication. Apparently it does not. It has been held to apply only to earlier proceedings although the relevant judgment has been subject to some criticism[50].

section 107(5)

Something being *in writing* includes it being recorded by any means.

section 107(6)

Experience is that when there is a disagreement as to whether or not there is a contract that is subject to statutory adjudication, the courts generally tend to look for reasons to say that one does exist rather than that it does not. There have however been some fine distinctions drawn.

It has been held that a written contract which was varied by an oral agreement was not a contract in writing[126]. In respect of an initial oral agreement between a consulting engineer and a contractor, it was held in the first instance that it was sufficient that general documentation demonstrated an agreement, the nature of the work and the price[95]. A similar view was taken in another case[12]. But in respect of the first case the Court of Appeal disagreed and decided, in the particular circumstances, that the substantial evidence in writing was insufficient to satisfy section 107(2)(c)

23

because it was not evidence of all of the items of the oral contract[96]. The approach of the Court of Appeal had to be followed in a later case[128]. However, the Court of Appeal also concluded that, if the agreement in writing contains all the terms material to the issues in dispute, the absence in writing of other terms which are immaterial to the issues may not prevent the dispute from being adjudicated[96].

2.5.3 Is yours a construction contract?

The Act applies only to construction contracts as defined in section 104 of the Act, but even so not all construction contracts are caught (exclusions are discussed later).

A construction contract for the purposes of the Act means an agreement with a person for any of the following:

    (a)   the carrying out of construction operations
    (b)   arranging for the carrying out of construction operations by others whether under subcontract or otherwise
    (c)   providing his own labour, or the labour of others, for the carrying out of construction operations.

section 104(1)

A construction contract also includes an agreement in relation to construction operations:

    (a)   to do architectural design, or surveying work,
    (b)   to provide advice on building, engineering, interior or exterior decoration or on the laying out of landscape.

section 104(2)

Thus section 104(2) ensures that all manner of service contracts are caught by the Act so long as they relate to construction operations. A claim for professional negligence may be adjudicated[47]. It is interesting to note however that the activity of a claims consultant in advising on an adjudication under the Act has been held not to be a construction operation[43].

It becomes apparent that whether or not a construction activity falls within the scope of the Act largely depends on what is meant by a *construction operation*. The term was introduced to enable a distinction to be made between various construction activities and to facilitate the exclusion from the Act of certain types of activity.

2.5.4   *What are construction operations?*

The work you are carrying out has to be a construction operation for yours to be a construction contract, but what actually are construction operations?

These are listed at section 105(1) of the Act. They are operations for:

(a)   the construction, alteration, repair, maintenance, extension, demolition or dismantling of buildings or structures forming or to form part of the land (whether permanent or not)

Also, similar activities in respect of

(b)   any works forming or to form part of the land including walls, roadworks, power-lines, telecommunication apparatus, aircraft runways, docks and harbours, railways, inland waterways, pipe-lines, reservoirs, water-mains, wells, sewers, industrial plant and installations for purposes of land drainage, coastal protection or defence

(c)   installation in any building or structure of fittings forming part of the land including systems of heating, lighting, air-conditioning, ventilation, power supply, drainage, sanitation, water supply or fire protection, or security or communications systems.

The expression *part of the land* is a term common in the field of property and its meaning is akin to that of fixtures and fittings in a domestic residence. Problems of interpretation have arisen. Shop-fitting items which could be removed, i.e. they were not fixtures, were held not to be part of the land[45]. And in a case concerning a module constructed in England for installation on an oil rig overseas, it was held that the eventual location had to be in England (or Wales) and not below low water line for the work to form part of the land[105].

(d)   external or internal cleaning of buildings or structures so far as carried out in the course of their construction, alteration, repair, extension or restoration

(e)   operations which form an integral part of or are preparatory to or are for rendering complete

any of the operations referred to at (a)–(d) including:

site clearance, earth-moving, excavation, tunnelling and boring, laying or foundations, erection, maintenance or dismantling of scaffolding, site restoration, landscaping and the provision of roadways and other access works

(f)   painting or decorating the internal or external surfaces of any building or structure.

2.5.5   *What are not construction operations?*

For reasons which are not entirely clear those responsible for drafting the Act were persuaded that certain areas of construction are different from the rest, particular considerations apply to them and they should be excluded from the requirements of the Act.

Section 105(2) lists operations that are not construction operations within the meaning of the Act:

(a)   drilling for or extraction of oil or natural gas
(b)   extraction (whether by underground or surface working) of minerals; tunnelling or boring, or the construction of underground works, for this purpose
(c)   assembly, installation or demolition of plant or machinery, or the erection or demolition of steelwork for the purposes of supporting or providing access to plant or machinery on a site where the primary activity is –
   (i)   nuclear processing, power generation or water or effluent treatment, or
   (ii)   the production, transmission, processing or bulk storage (other than warehousing) of chemicals, pharmaceuticals, oil, gas, steel or food and drink
(d)   manufacture and delivery to site of –
   (i)   building or engineering components or equipment
   (ii)   materials, plant or machinery, or
   (iii)   components for systems of heating, lighting, air conditioning, ventilation, power supply, drainage, sanitation, water supply or fire protection, or for security or communications systems.

However, if a contract is for supply and fix then all at (d) become construction operations. And if a hire agreement includes the provision of an operator with the machine, this becomes a construction operation[123].

(e)   the making, installation and repair of any artistic works–sculptures, murals and other works which are wholly artistic in nature.

Primary activity at (c) relates to the site as a whole, not to a particular part where the work subject of the dispute is being executed[2].

Whilst section 105(1) in itself has not caused any difficulty of interpretation, when read with 105(2) a number of problems have arisen which have required resolution by the courts.

Section 105(1)(e) states that scaffolding operations *preparatory to such operations as are previously described in this subsection*, i.e. 105(1)(a)–(d) are construction operations. It may be thought therefore that scaffolding for operations at 105(2)(c), i.e. operations which are defined as not being construction operations, would consequently not be a construction operation. However, notwithstanding the words *where the primary activity is*, the court has held that it is[85].

In contrast, cladding and insulation to pipework of a 105(2)(c)(ii) operation have been held in a particular instance to follow the classification of the work of which the pipework was part, i.e. it was not a construction operation[3]. Likewise the assembly and installation of boiler plants on a site adjacent to an oil refinery were also excluded as the plants supplied steam to the refinery and therefore furthered the primary activity of the site[74].

Whether or not pipework itself is a construction operation depends on its function. If it connects plant which is excluded under 105(2), it is considered as part of the excluded plant[56].

Of importance is the decision that *is* in the expression *where the primary activity is* at 105(2)(c) also means 'will be'[3].

2.5.6   What if your contract is mixed?

You may have a mixed contract – part construction operations as defined and part not. The Act applies only to the former. You cannot claim that it applies to your contract as a whole on the basis that it applies to part of it[56].

section 104(5)

2.5.7   Where is your contract being carried out?

The HGCRA applies only to construction contracts which relate to the carrying out of construction operations in England, Wales or Scotland[105].

section 104 (6)(b)

If your contract is in Northern Ireland, the Construction Contracts (Northern Ireland) Order 1997 applies.

2.5.8   Exclusions

In addition to those construction operations excluded by virtue of section 105(2), there are certain types of work to which the Act does not apply.

2.5.8.1   Residential premises

If your contract relates principally to operations on a dwelling and is with a residential occupier, that is someone who occupies or intends to occupy it as a resident, then the contract is excluded from the Act[129].

What is a *dwelling*? It is a dwelling house or flat but a dwelling house does not include a building containing a flat.

What is a *flat*? It is separate and self-contained premises constructed or adapted for use for residential purposes and forming part of a building divided horizontally from the rest.

sections 106(1)(a) and 106(2)

It is odd that, for the purposes of the Act, a vertical division of a building will not apparently create a flat.

Difficulties may arise if your construction operations within a single contract are partly for a residential occupier and partly not; that is, how should *principally* at section 106(2) be interpreted? This has not been positively established but in a particular situation the court has stated that 65% was not sufficient for the contract to be caught by the Act[97].

To add to the confusion the court has decided that an annual service and maintenance contract for heating systems, gas, fires and works within residential properties is a construction contract[80].

2.5.8.2   Other

The Act does not apply to a construction contract entered into by or on behalf of Her Majesty in her private capacity. However it does apply to a construction contract entered into by Her Majesty on behalf of the Crown. The Act also applies to any construction contract entered into on behalf of the Duchy of Cornwall.

section 117(1) and (2)

Certain types of contract are excluded from the operation of the Act. The Act does not apply to *any other description of construction contracts excluded from the operation of this Part by order of the Secretary of State*. Such an order is The Construction Contracts (England and Wales) Exclusion Order 1998.

section 106(1)(b)

There are five particular types of contract that are excluded by the Order:

(1)   contracts relating to agreements made under specified statutory instruments, in particular those relating to:

- Sections 38 and 278 of the Highways Act 1980 as partly amended by or substituted by the New Rules and Street Works Act 1991
- Sections 106, 106A and 299A of the Town and Country Planning Act 1990 as partly amended by and added to by the Planning and Compensation Act 1991
- Section 104 of the Water Industry Act 1991

(2) externally financed development agreements within the meaning of section 1 of the National Health Service (Private Finance) Act 1997

(3) contracts entered into by certain identified bodies under the private finance initiative as defined

(4) contracts that are finance agreements as defined

(5) contracts that are development agreements as defined.

To summarise, there are certain construction contracts which are excluded from the requirements of the Act. Of those which might otherwise be included some are not caught because they are not construction operations as defined.

So there may be some difficulty in deciding whether or not your contract is subject to the Act. Consider it critically and objectively rather than hopefully. If you are still of the opinion that it is not a construction contract that is excluded and it is a construction operation, then go ahead. The other party will have to satisfy the adjudicator that your contract is excluded in some way and that he does not therefore have the jurisdiction to hear the dispute. If he does not agree with them and proceeds, they could request the courts to intervene and stop the adjudication, or wait and, if the decision goes against them, refuse to comply and then resist enforcement.

2.6 Other considerations

2.6.1 *Does the law of the contract make any difference?*

The party for which you are carrying out work may be a foreign organisation and for some reason it has designated that the law of your contract is not that of England, Wales or Scotland. The relevant law of the contract does not matter – the Act still applies.

section 104(7)

The same principle applies to construction work subject to Northern Ireland legislation.

2.6.2 *What if your dispute is already the subject of some other dispute resolution or legal process?*

You can instigate adjudication even though you are already engaged in some other dispute resolution process in respect of the same dispute[52/53/91]. The other process is unaffected and will continue. That said, it would perhaps not be a good move if you are engaged in mediation or conciliation.

An arrestment – the protective measure available in Scotland whereby monies that are the subject of a dispute may be frozen pending resolution of the dispute – has been the subject of attention by the courts there.

It has been held that whilst an arrestment cannot be used to avoid the consequences of an adjudicator's decision[94], a party cannot insist on recall of an arrestment against it following a decision in its favour; the other party can require that the arrestment remains in place until final resolution of the dispute[106].

2.6.3 *What if your contract has been allegedly repudiated or has been terminated?*

You may still refer a dispute to adjudication even though your contract has been repudiated or allegedly repudiated; indeed, it has been held that repudiatory breach is a matter within the jurisdiction of an adjudicator[79].

Even if your contract has been terminated or determined by yourself or the other party, the right to adjudicate survives and you can still utilise it[1].

2.6.4 *Can you adjudicate on more than one dispute at the same time?*

There is no limit to the number of disputes you may refer at one time but there may be restrictions as to the number arising out of the same contract that you can refer to the same adjudicator. The Act is silent on the matter so you have to look at the adjudication provisions of your contract in order to determine the position. Most are not restrictive but the Scheme, for example, states at paragraph 8(1) that an adjudicator can adjudicate on no more than one dispute under the same contract at a time unless the parties agree otherwise.

When put to the test the Scheme restriction was upheld[50]. But whether more than one dispute has been referred depends on the meaning of the word *dispute*.

The TeCSA Rules give clarification on this point. Paragraph 13 of the 2002 version states that:

> *The scope of the adjudication shall be the matters identified in the notice requiring adjudication . . .*

This definition was accepted literally in a case where the TeCSA Rules applied[8].

As with the Scheme, other matters may be decided if the parties agree. Additionally under the TeCSA Rules, paragraph 13(ii), the adjudicator may decide matters which he determines must be included *in order that the adjudication may be effective and/or meaningful.*

But under other procedures interpretation by the courts has been variable. There was some criticism[91] of an early rather wide general interpretation that a party could choose to include in an adjudication reference all or any of the disputes that then existed[42]. A tighter approach was subsequently adopted by the courts but it was acknowledged that one dispute could contain any number of issues[11/27/71].

It may be convenient and economical for the same adjudicator to handle related disputes under different contracts – e.g. a main contract and subcontract under the same project – but again, the Scheme at paragraph 8(2) precludes this without the consent of all the parties involved.

You need to check the adjudication provisions in your contract in order to ascertain whether they contain any such restrictions.

2.6.5 Duplication of referrals

If your dispute has already been referred to adjudication or is substantially the same as a dispute that has previously been referred, you are not allowed a second try; even though you may have secured an appointment the adjudicator must stand down as soon as the duplication is established[76/99].

The Scheme at paragraph 9(2) explicitly states this and the Act impliedly does so at section 108(3) – the decision of the adjudicator, that is the first decision, is binding on the parties until finally determined by legal proceedings, by arbitration or by agreement. If a second adjudicator is appointed and makes a decision on the same dispute it will be set aside. Where a second referral under the same contract is valid, the adjudicator cannot revise or vary the decision of the first[113].

However, a court has held that where the relief claimed in a second adjudication was different to that in the first, the adjudicator was correct in assuming he had jurisdiction and proceeding to a decision even though the second dispute arose from the same circumstances[55]. The same approach was taken where, although the same facts were argued, a second adjustment related to a final account whereas the first concerned an interim amount[102]. And a contractor who, having failed in an adjudication because of lack of a withholding notice, then himself raised a second adjudication concerning the value of the subcontractor's work, was permitted to proceed with adjudication[133].

In contrast, where a party which had been ordered at adjudication to pay the full amount applied for because there had been no proper withholding notice against the application, then referred to a second adjudicator the question of what was the properly calculated sum due under the application, the second referral was held to be invalid as the dispute had been considered and decided by the first adjudicator[115].

2.6.6 Settlement

You cannot refer a dispute in respect of which there has been a binding settlement; and the settlement agreement itself is not a construction contract subject to statutory adjudication[63]. A settlement stands on its own and you cannot use adjudication to open up its terms even if you consider you were under some duress at the time to accept terms that were unreasonable. Whether the agreement was invalid on account of the alleged duress is a matter to be decided outside the adjudication[98].

It is considered that a settlement removes any disputes which existed and if the agreement included an undertaking that you would not in the future refer any matter to adjudication, then you cannot rely on the Act to override that undertaking[44].

2.6.7 Can the responding party refuse or delay adjudication?

The entitlement to refer a dispute to adjudication at any time is a statutory right which cannot be removed even by an ad hoc agreement between the parties[28]. There is no mechanism whereby the other party can apply for a deferment of the proceedings – only you could agree to that – and the other party refuses to participate at its peril.

Your contract may specify some mandatory alternative dispute resolution process which has to be followed before a dispute can be referred to arbitration or litigation. The requirements of the Act override any such provision and the responding party cannot insist on it being pursued prior to the commencement of adjudication[25].

The court has criticised the concept under certain forms of contract (ICE and ECC) that no dispute arises until a *matter of dissatisfaction* has been investigated – or not investigated upon request[75]. This preliminary step is discussed at Part 6 of this book. However, the comments of the judge were not actually part of his judgment (he was not asked to decide the point) and so, although persuasive, did not become law.

Despite considerable initial support for the view expressed by the judge, the drafters of the two forms of contract have made no effort to remove the restriction. Interestingly a more recent case, in which the nature of the matter in dispute was considered[82], has led to greater support for the view that the concept of matter of dissatisfaction is not an invalid restriction.

2.6.8 What if the responding party is in financial difficulties?

Obviously, before you commence adjudication proceedings, careful consideration needs to be given if the responding party is in a dubious financial state. Is it worthwhile embarking on the cost and effort of an adjudication if they may not be able to pay anything you are awarded? If the responding party is struggling financially and could be driven into bankruptcy if a decision in your favour is enforced, you should be aware that you may have difficulty securing enforcement. The financial state of the responding party at the time of enforcement proceedings is considered later in this Part, at sections 8.2 and 8.3.

Even though the prospect of obtaining any money from the other party is remote or even non-existent, you may have some reason for wishing to pursue a claim against them.

However, if the other party is so seriously in trouble as to be insolvent, you will first have to overcome a legal hurdle. Adjudication has been held to be legal or other proceedings for the purposes of Section 11(3) of the Insolvency Act 1986[108], which prevents action against a party that is the subject of an Administrative Order without the leave of the court or the consent of the administrator[24].

2.7 *Summary*

It is important that you satisfy yourself that you can properly proceed to adjudication. If your referral is held to be invalid then you will be liable for all of the costs that the adjudicator has incurred.

So before deciding that you will proceed to adjudication check that:

(1) It is the right course of action
(2) You have a genuine referable dispute
(3) You have a contract and it does not predate 1 May 1998
(4) Yours is a construction contract as defined at section 104 and it is in writing
(5) The construction operation under which the dispute arises is one of those included under section 105(1) but not excluded under section 105(2)
(6) Your contract is not excluded from the Act:
    (a) because it concerns work at a dwelling as defined at section 106 and you or the other party is the residential occupier
    (b) by virtue of section 117(1) of the Act or of the Exclusion Order 1998.

3. *What next?*

3.1 Which adjudication procedure?

Most standard forms of contract now have adjudication provisions which will be Act compliant. But check that you are not working under a version of a standard form produced before statutory adjudication was introduced; although the current edition of the form will likely contain the minimum requirements of the Act, your version will not and the Scheme will therefore apply.

The Act simply states certain minimum requirements that a construction contract must contain. Something else is needed in order that the intentions of the Act can be sensibly implemented. Adjudication procedures are the operation manuals of the Act.

If your form of contract or agreement has been produced since statutory adjudication was introduced, it will almost certainly embody a full adjudication procedure; alternatively the adjudication provision itself may be limited but it will incorporate by reference one of the standard procedures that are available.

If you are an employer's representative the adjudication provisions in your terms of engagement will normally be a slightly modified CIC Model Adjudication Procedure. As a main contractor the adjudication procedure will be the one related to the standard form of contract under which you are probably working, although some employers have their own procedure. If you are a sub-contractor you may well be under the main contractor's own form of contract including his own adjudication procedure. Should no particular procedure be stipulated the Scheme will apply unless you and the party with whom you are in contract agree to use some other procedure.

Whilst, in order to conform with the requirements of the Act, most standard adjudication procedures (whether within the conditions of contract or separate) follow a similar general pattern, each may differ in its particular requirements. So it is necessary to study your procedure carefully to ensure that you are aware of what you have to do. For example, the provisions in the ICE Conditions of Contract contain the concept of a matter of dissatisfaction which has to be referred to the engineer before a dispute referable to adjudication can even arise.

Remember that if your procedure does not fully comply with the requirements of section 108 of the Act, it is supplanted by the Scheme. It is not possible to retain bits and pieces of a non-compliant procedure and simply replace the defective parts with extracts from the Scheme.

3.2 Do you need to engage a lawyer or consultant?

If a party considers it has a referable dispute its arguments ought already to have been comprehensively presented to the other party in a manner that enabled proper consideration to be given to them. In that event the party ought to be able to present its case to the adjudicator.

However, despite hopes that parties to adjudication would refrain from engaging lawyers or consultants, all too often outsiders are brought in to prepare and present submissions. Such practice may be a matter of policy in your organisation. But perhaps because of a lack of particular skills or a limitation on resources that can be diverted from other ongoing activities, you may feel you should obtain assistance of some sort.

When considering whether or not to call in outside expertise acknowledge your limitations and do not proceed alone if it is possible you will be out of your depth. Thus if you have any doubts regarding legal aspects or contractual interpretation you would be wise not to proceed unaided.

If you decide to obtain assistance be careful in your choice.

Do not necessarily engage your usual solicitors. They may be excellent at attending to your routine company affairs but may know nothing about the construction process, much less about adjudication. On the other hand it is not necessary to go to one of the top legal firms unless you have plenty of money and you do not mind expending unrecoverable large sums on their fees. If you do not already know of a suitable firm of solicitors ask around and select one which has already provided good service to someone else.

The same is true of consultants, a number of whom are extremely expensive. Again, there are many smaller firms or practices who will provide an excellent service at reasonable cost, but do get a reference before making your choice.

Whether you choose a solicitor or consultant, first ascertain their scale of charges. Some, particularly consultants, may offer to prepare your referral notice for a reasonable sum but will then charge extortionately for the subsequent work which will inevitably be necessary and will frequently be substantial. A rate per hour may be reasonable but you can find in due course that the hours allegedly expended are high for the work that has been carried out. Before you commit yourself ask for an estimate of the overall time that it is expected will be necessary. If you have doubts investigate an alternative adviser; there is plenty of competition.

4. *What will it cost?*

Costs fall into two categories:

- Those of the adjudicator
- Those of each of the parties.

The Act itself is silent on both categories.

4.1 The adjudicator's costs

The Scheme and most standard adjudication procedures permit the adjudicator to allocate his own costs between the parties as he deems appropriate.

If there is a relatively clear winner and loser it is to be expected that an adjudicator will order that the unsuccessful party pays his costs. If the claim has succeeded only in part he may direct that his costs are shared between the parties.

You should be aware however that the parties are what is known as jointly and severally liable for the adjudicator's costs. This means that if the responding party has been ordered to pay all or part of the adjudicator's costs and does not do so, the adjudicator can call on you to make up the deficit and you will then have to take action to recover this from the responding party.

The adjudicator's costs will be difficult to predict as there are a number of influencing factors, as follows.

4.1.1   The adjudicator's hourly rate

This, at the time of writing, can vary from about £80 to perhaps £300. The least expensive are less experienced engineers, architects, quantity surveyors and the like. The more experienced construction professionals will probably charge between £100 and £175. If you ask that your adjudicator is legally qualified – as a number of construction professionals are – you will pay at the top end of this range or above. If a solicitor or barrister is selected then you will find yourself paying the highest rates unless the TeCSA Adjudication Rules are operative, in which case the adjudicator is presently limited to £1250 per day.

4.1.2   The size and complexity of the dispute

Obviously the adjudicator will need to spend more time on a more involved or difficult dispute. Additionally, the behaviour of the parties will influence the hours he has to expend. Inadequately prepared submissions, failure to react to directions, the raising of challenges and so on all add to the adjudicator's workload. If he has to convene a meeting or visit site this can add significantly to both the hours expended and expenses incurred. Try to sort out any problems directly with the other party rather than passing them through the adjudicator.

4.1.3   The ability and experience of the adjudicator

Notwithstanding that they may have received adequate training, less experienced adjudicators will expend more time pondering procedural matters than those who have conducted many adjudications; and the more peripheral the nature of the dispute to his own practical experience, the greater the time the adjudicator will devote to deliberating the technical aspects of the issues before him. All this apart, some adjudicators just work faster than others. It must be acknowledged that the performance of adjudicators does vary and, to a degree, you must accept that this will be reflected in the time they will take to reach a decision. If there is such a thing as the average adjudication, a reasonably competent adjudicator would expend 30–40 hours in reaching his decision. Jurisdictional complications will increase this range.

Adjudicators are entitled to obtain either technical or legal advice. Although properly trained adjudicators have a knowledge of the law they are not lawyers and the less experienced adjudicators particularly will tend to obtain legal advice if some procedural problem or point of law arises. It should be only rarely that an adjudicator selected for his practical experience needs to seek expert technical advice. If however the adjudicator does obtain help, then unless it can be shown that it was not appropriate for him to do so, it is the parties who have to pay for it.

An adjudicator is entitled to receive reasonable payment for carrying out his duties. The courts are reluctant to interfere in the matter of what are reasonable costs and will do so only in extreme cases or where there has been bad faith on the part of the adjudicator[58/109].

4.2   The parties' costs

At the time of the inception of statutory adjudication it was envisaged that the type of dispute that would be referred would be relatively simple, and parties were encouraged to represent themselves. As is usual, things have not worked out quite as expected and adjudications are commonly quite complicated. The result is that parties frequently turn to lawyers or consultants to handle matters for them and consequently the cost of referring a dispute to adjudication can be significant – proportionately more so if the amount in dispute is small but the arguments complex.

The position regarding the authority of the adjudicator to allocate the parties' costs varies according to which adjudication provisions are operative. The ICE, IChemE and JCT procedures, for example, state that each party must meet its own costs, but GC/Works/1, the form generally used on government contracts, actually requires the adjudicator to state how the parties' costs are to be allocated. The

Scheme and some procedures are silent on the matter. The fear of having to pay the other party's costs as well as your own if things go wrong could be a deterrent to referring a dispute to adjudication. Hence any rule that each party must meet its own costs regardless of the outcome of the adjudication is generally welcome.

Since the inception of statutory adjudication there has been speculation as to whether, when the procedure is silent, an adjudicator can allocate costs. There were two early contradictory court decisions in situations where the Scheme applied. The first[33] found the adjudicator does have authority; the second, and generally the preferred judgment[79], was that he does not. More recently a third judgment, following the preferred judgment, emphasised that costs could not be recovered as damages[110].

If you are under a procedure which is silent on the matter of parties' costs but you are supremely confident of winning – so much so that you consider the risk of having to pay the other party's costs negligible – you may be able to persuade them to enter into an agreement giving the adjudicator the power to allocate costs[4/77/79]. Such agreement would need to be in writing and the adjudicator made aware of it before he makes his decision. If your adjudication provisions are those of, or for use with, one of the standard forms, your agreement will need to state clearly that any clause stipulating that each party bears its own costs (for example clause 41A, 5.7 in the case of JCT Standard Building Forms) is withdrawn by consent.

Notwithstanding that the parties have not agreed that the adjudicator may allocate their costs, they may inadvertently give him jurisdiction to do so if, in their submissions, each seeks to recover its costs from the other and requests the adjudicator to order accordingly[77/79].

If the adjudicator has jurisdiction to allocate parties' costs he can, if requested, also assess the amount of them[27].

4.3 Special provisions as to costs

Some bespoke adjudication procedures (primarily those prepared by contractors) stipulate that the referring party shall meet its own, the other party's and the adjudicator's costs regardless of the outcome of the adjudication. Such provision is obviously intended to deter subcontractors and was clearly not contemplated by those who drafted the Act. Nevertheless, if you have been successful in your adjudication but such a clause exists in your contract, you must at the present time either accept it or try to persuade the court that it is oppressive, contrary to the spirit of the Act and should be set aside. Since there is already a judgment that such a clause, however objectionable, is valid[19] you would have a difficult task.

5. *Getting the adjudication under way*

5.1 With whom or to where do you correspond?

If your contract states an address for the service of notices. that is where you write to in the first instance. If reference is made to the principal place of business that means head office as opposed to a regional office. If the address is different from that to which you normally write, then copy your notice of adjudication to the usual place. You should then be advised by the other party as to where to direct future correspondence, possibly to a legal or other representative. If you are not sure, you should at the earliest opportunity ask for confirmation as to where communications are to be sent.

If you are a contractor and your dispute arises under the contract with the client or employer, it is with him that you are in dispute not the architect or the engineer who has been administering the contract. Your notice should therefore be sent to the client but copied to the engineer, architect or the project manager as the case may be. Almost inevitably you will be requested to direct future correspondence to the person who has been administrating the contract unless a solicitor or consultant has been engaged.

If you as client are instigating adjudication proceedings, it is in order for your architect or engineer to send the notice of adjudication and to handle the proceedings on your behalf. If you engage a different agent, solicitor or consultant, then you or he must confirm to the other party that he acts with your authority.

Sometimes a principal engages an agent to organise and administer the project entirely, including the placing of work contracts on behalf of the principal. The law and rules of agency apply. The contracts between the principal and the work contractors arranged on his behalf by his agent are still construction contracts to which the Act may apply even though the principal himself has had no direct contact with the works contractors[112].

Check your adjudication procedure to see if the method of communication is stipulated. Whilst, due to the time constraints, adjudications are normally conducted by facsimile (although there is an increasing use of e-mail), you may be required to send notices and other submissions by some particular method. For example, the

JCT Procedure requires that if the referral notice is given by facsimile it must forthwith be sent by first class post or given by actual delivery.

The adjudicator may be required to confirm to the parties the date of receipt of the referral, but it is good practice to ask him to acknowledge receipt of any submission you make. Your fax machine may record a successful transmission but that is not to say that the adjudicator himself received it. The adjudicator should, as appropriate, react promptly to any communication you send. If he does not and it is important, ask for confirmation of receipt.

5.2   The notice of adjudication

In order to get adjudication under way you must first advise the other party of your intention by giving it what is called a notice of adjudication. This must be in writing and, to avoid possible arguments later, should be sent recorded delivery or similar. Indeed, certain procedural rules require notices to be sent by recorded delivery.

The notice must contain certain information. Exactly what this is depends on the applicable procedure; the requirements of the Scheme at paragraph 1(3) are one of the most detailed:

(a)   The nature and a brief description of the dispute and of the parties involved
(b)   Detail of where and when the dispute has arisen
(c)   The nature of the redress which is sought
(d)   The names and addresses of the parties to the contract (including, where appropriate, the addresses which the parties have specified for the giving of notices).

Whilst other procedures may require less information it is absolutely essential that you clearly state what it is you wish to refer. In the early days of statutory adjudication more emphasis was placed on the referral notice but it has subsequently been acknowledged that the notice of adjudication defines the dispute to be referred[72]. So do not be misled by terminology such as, for example, *briefly identify* in the JCT adjudication provisions or *a brief description* in the Scheme. Be brief but be specific.

Adjudication proceedings have been stayed and decisions not enforced by the courts because of uncertainty as to what was being referred. Where a notice merely referred to correspondence without extracting and restating the relevant parts, the issues so affected were held to have been invalidly referred because it was not possible to identify what was actually being referred; the adjudicator consequently lacked jurisdiction except in respect of the identifiable issue[49]. On the other hand there have been what some commentators have considered to be rather generous interpretations as to whether disputes have been properly referred[59]. Nevertheless, you should not assume that the courts will overlook inadequacies in a notice of adjudication.

The procedural rules under which you are operating will state whether there are any restrictions on the number of disputes you can refer to the same adjudicator at one time. The Scheme states only one unless the parties agree otherwise. However any such restriction need not be as limiting as it may first appear.

Your dispute may contain any number of issues, for example it may be in respect of the amount of an interim certificate. Your referral can challenge any of the elements which comprise or influence the amount of the certificate – measurement, variations, loss and expense, and so on – and the validity of a withholding notice[62]. Indeed the Scheme restriction that the adjudicator can decide only one dispute at a time is generally interpreted quite loosely (see earlier in this Part, at section 2.6.4).

If, as is generally the case, you are seeking money, it has been held that it is not necessary for you to specify the amount[42]. Although not the best approach, your notice may simply ask the adjudicator to decide what sum is due – subject, of course, to your having established that a dispute exists.

Provided the intentions of the referring party are made clear in the notice of adjudication and the subsequent referral, experience is that, normally, the other party will not object and the courts, if asked, will tend to be supportive. It is clearly more convenient to deal with everything at one time – since the facts and evidence are frequently common – rather than have a series of separate adjudications. What is important is that the matters you are referring have been considered by the responding party and rejected or that they have been ignored.

Do not rush into sending off your notice of adjudication – depending on the operative adjudication procedure, time could start to run from the date that you send it. Late referral to the adjudicator may not be a disaster but to avoid a possible challenge it is prudent to make sure your referral notice is on time. To this end ensure that it is ready or is essentially complete before you send off your notice of adjudication.

5.3 Getting an adjudicator in place

The process is termed naming, nominating, appointing or selecting the adjudicator. He may be:

(1) Named in the contract
(2) Agreed by the parties and appointed by them
(3) Nominated, appointed or selected by an adjudicator nominating body.

5.3.1 Named in the contract

This is a practice which, with rare exceptions, is confined to main or principal construction contracts. Although it has certain disadvantages, its adoption avoids the scramble to get an adjudicator in place every time a dispute arises; he is there and ready to act when called upon.

It is possible that several persons of different disciplines (e.g. surveyor, civil engineer, architect) may be named in the contract and you can approach whoever is most appropriate. If one or more potential adjudicators are named you will, by entering into the contract, be deemed to have accepted that any one of them may act if required. If you are an employer you should have inserted into the contract only those names with which you are content. If you are a contractor you should have ensured that your tender or acceptance did not include the names of any persons whom you would not wish to act as adjudicator.

Where an adjudicator is named it is possible that the contract will also include a procedure to be adopted when calling upon him to act. Be careful to ensure that you comply with any such procedure.

5.3.2 Agreed between the parties

Whilst all listed adjudicators should have gone through a process of training and assessment, it must be acknowledged that the quality of some is less than it should be. Consequently, if the adjudicator is not prenamed, it is preferable to try to secure someone who is not only technically suitable to handle the dispute in question but who is also an experienced adjudicator. However, you cannot impose a particular adjudicator on the other party; you must obtain its agreement.

Either party may suggest who should act as adjudicator. As the referring party you are clearly best placed to make a first proposal

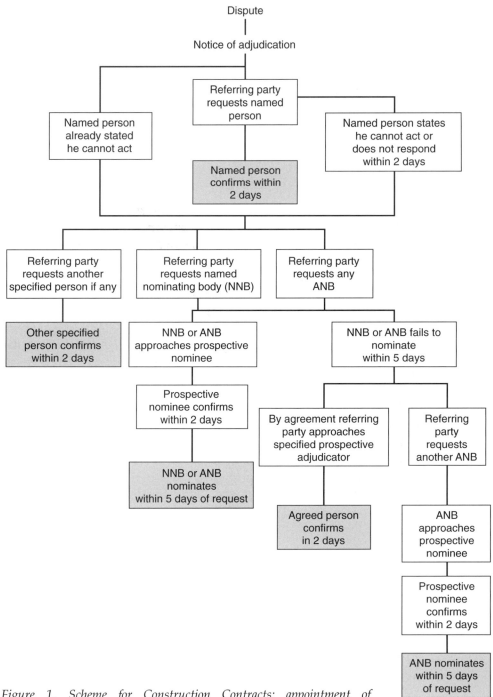

Figure 1. Scheme for Construction Contracts: appointment of adjudicator when adjudicator is named in contract.

but before doing so you will of course need to be sure that the person is willing and able to act.

You must only propose someone who is independent of you and your company or partnership and who, it would be evident to an outsider, would not be biased. Your proposal may be made with the notice of adjudication but this is not essential. The other party, not being obliged to accept your suggestion, may offer an alternative name which you may accept or reject. To improve the chances of reaching agreement, you may offer several names from which the other party can make a choice.

If you are under a contract that embodies a standard form of adjudicator's agreement, then anyone you approach must be advised accordingly and of any amendments that have been made to it.

You will want to get the adjudication under way as quickly as possible so do not spend too much time trying to agree with the other party who is to be the adjudicator.

5.3.3 *Nominated by an adjudicator nominating body*

If agreement is not possible (and you are not obliged to try to get agreement unless the adjudication procedure says so and none of the standard procedures do) you must apply to the adjudicator nominating body (ANB) stipulated in your contract. If none is stipulated you may go to any one of a number of ANBs but the obvious choice is the one most closely related to the subject matter of the dispute. A list of the principal ANBs appears in Appendix 6 at the back of this book.

An adjudicator nominating body is defined in the Scheme, paragraph 2(3) as:

> a body (not being a natural person and not being a party to the dispute) which holds itself out publicly as a body which will select an adjudicator when requested to do so by a referring party.

Not surprisingly the standard required by ANBs of those on their lists varies considerably and this has given rise to some concern. ANBs are not regulated although some influence is exerted by the Construction Industry Council over those professional institutions under its umbrella. If therefore an ANB is not stipulated in your contract, choose one of the professional institutions and you will stand the best chance of getting a competent and experienced adjudicator. Otherwise you should enquire of an ANB what they require of those on their list as regards experience and training, both basic and continuing.

You first need to obtain an application form for the appointment of an adjudicator. A telephone call to the ANB will secure this. All charge a fee to be paid by the party making the application. The nomination procedure is fairly standard and will be explained in the documents sent to you. Inevitably you will need to provide details of the parties, the contract and the nature of the dispute to assist the ANB in making a suitable nomination. You will be asked to state the expertise you consider that the adjudicator should have but ANBs do not guarantee to appoint a person who fulfils your requirements.

From the time you make a request the ANB normally has only five days in which to make a nomination. Remember that your referral notice has to be sent to the adjudicator within seven days of your notice of adjudication, so it is essential that you complete quickly and accurately the documents sent to you and return them promptly. Communication is normally by fax but most ANBs accept e-mail.

There is a tendency, because at the end of the day the vast majority of referring parties are seeking money, to assume that the correct choice of adjudicator is a quantity surveyor. If the issues revolve around measurement and bills of quantities that expertise is most appropriate, but you should, when completing your application, consider the grounds on which the claimed entitlement is based.

Perhaps there were changes in design, difficulties in the foundations or structure, interference with the programme. All these are matters which an engineer or architect may be better suited to decide. On the other hand, if your dispute is about interpretation of the contract, conditions of engagement or the like, then perhaps a lawyer would be more appropriate.

You should be able to secure an adjudicator of the profession of your choice but you may not get one with particular technical experience. An ANB will do its best but it does not undertake to provide an adjudicator with specialist experience.

If you have no intention of trying to agree an adjudicator with the other party then you may be able to save time by obtaining the application documents from the ANB before you give your notice of adjudication. These can then be completed and returned on the same day as you give your notice of adjudication. However, some ANBs require a copy of the notice with your application.

Having secured the agreement of the adjudicator to act, the ANB

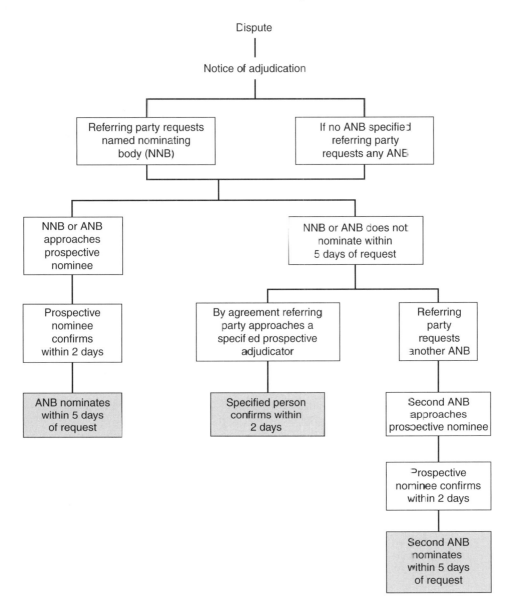

Figure 2.   Scheme for Construction Contracts: appointment of adjudicator when adjudicator is not named in contract.

will communicate his details to the other party as well as yourself. Neither you nor the other party can object to the person selected unless he has had some connection or association with one of the parties that might lead to bias or partiality, or unless you are seriously and justifiably concerned as to his suitability to deal with your dispute. In either case he ought not to have accepted the appointment.

If you are genuinely concerned regarding the suitability of the appointed adjudicator you should seek the support of the responding party; if it shares your concern a joint approach is likely to be more effective. Possible action is discussed at Part 4, section 1.2, in this book.

5.4 The Adjudicator's Contract, Agreement or terms and conditions

A number of standard forms of contract include as part of their adjudication provisions, either directly or through an Adjudication Procedure, an Adjudicator's Contract or Agreement. This is a three-way agreement between the adjudicator and the parties and is inevitably a fairly basic and innocuous document containing none of the unattractive items that some adjudicators include in their own terms and conditions. In entering into a contract embodying an Adjudicator's Agreement, the parties also agree to accept the terms of that agreement. The adjudicator cannot vary his terms without the consent of both parties. If there has been any amendment to a standard agreement then the ANB should be informed of this at the time it is requested to appoint, so that a potential adjudicator can be advised before he accepts the appointment.

If the adjudicator is named in the contract his hourly charging rate will be stated in the Adjudicator's Agreement, which may be signed at commencement of the contract or when he is first called upon to act.

In situations where the adjudicator is appointed on demand rather than prenamed in the contract, the parties will still have agreed to the terms of any Adjudicator's Agreement which the contract contains or incorporates. When he is appointed the adjudicator automatically accepts the terms of the agreement. At that time he will insert his hourly charging rate. In some standard forms, the JCT family of contracts for example, there is an alternative provision for the adjudicator to state a lump sum rather than an hourly rate. It is often not realised that the parties are not obliged to accept the rate or sum inserted as this is something that has arisen after the subject contract had been entered into. If you consider his rate is unreasonably high then you may object. An adjudicator is entitled only to reasonable payment for his services but what constitutes a reasonable hourly rate is open to interpretation and will reflect his normal professional charging rate. Thus you can expect to pay more for a lawyer adjudicator than for an engineer. It is important that you register any objection at the time, i.e. you give a qualified acceptance of the Adjudicator's Agreement or Contract. Otherwise you may be held to have accepted the rate.

If you seek to agree an adjudicator with the other party, the adjudicator is in a position to require the agreement of you and the other party to his terms and conditions before he agrees to act. If either of you do not like these you may propose adjustments which he is asked to accept. Depending on the outcome he will or will not agree to act and, if he does not, you will obviously have to look elsewhere.

If the adjudicator is appointed by an ANB he is not entitled to insist on acceptance of his own terms and conditions although many adjudicators, upon appointment, do write to the parties requesting their agreement to these. You should read them very carefully as they sometimes contain quite unreasonable terms to which you should not agree. Some ANBs provide guidelines to their listed adjudicators advising them of the sort of terms that are and are not acceptable.

If you have any objections to the terms and conditions you should state them immediately; failure to do so may be construed at a later date as acceptance on your part even though you have not actually signed them. On the other hand if the adjudicator proceeds after you have registered an objection or you have specifically refused to sign (giving your reasons) and you have stated that you are continuing with the adjudication without prejudice, then he will be deemed to have accepted your amendments, or his terms and conditions will be of no consequence.

It is a good idea to try and work with the responding party. There is always a niggling feeling that if you alone have raised objections the adjudicator may, albeit subconsciously, favour the responding party in some way. He should not, but it can happen.

If the appointed adjudicator should refuse to act until his terms and conditions are accepted by the parties (and under the Scheme, paragraph 2(2) he must advise the parties within two days of being nominated that he has agreed to act) then you can revert to the ANB and request another nomination.

5.5 What happens if there is a delay in the appointment of an adjudicator?

If you do not give the ANB sufficient time to make a nomination or if the ANB itself is dilatory or has encountered a problem in finding a suitable person, you may find that an adjudicator is not in place within the stipulated time – often five days after the ANB has received a request. Whilst a late nomination is undesirable it need not, it is suggested, in itself be critical. There may be a consequential delay to the date of referral and this is considered later in this Part, section 5.7.

If nomination is late you should send off your referral notice immediately you know who the adjudicator is and explain the circumstances to him (although he should already be aware from the date of documents sent to him by the ANB). Except in the unlikely event that the other party has been prejudiced by the delay, the adjudicator should accept the situation. If he does not the worst that can happen is that he may declare that in the circumstances he cannot adjudicate and you will have to obtain another appointment. In that event, if the ANB was at fault they should not charge you for the second appointment.

5.6 The referral notice

5.6.1 *General considerations*

The referral notice will normally be required to be submitted within seven days of the notice of adjudication and is sent to the adjudicator and copied to the responding party. The name referral notice is misleading; as a whole it is more than a notice – it is your claim submission.

You need to be precise in the referral notice, just as you did in the notice of adjudication. The responding party is entitled to know exactly what the dispute is that is being referred. And remember, you cannot include more issues in the referral notice than you did in the notice of adjudication; the latter is now generally considered to be the key document determining what the adjudicator has jurisdiction to decide.

You must not take the view, simply because you may have been arguing the dispute for months, that the responding party should know what you are talking about. In law the onus of proof is on the person making an assertion. It is not good enough, for example, just to state that you were delayed in carrying out your obligations and what it cost you as a result or what extra time and resources you expended in administering the contract. Proper explanation and justification is required.

A vital point that many referring parties overlook is that it is now the adjudicator that they have to convince, not the other party. Failure to provide a comprehensive and properly supported referral submission that is understandable by a stranger to your contract will cause delay and extra expense and, most importantly, may result in you failing to convince the adjudicator.

This does not mean that you can now introduce new arguments not previously put to the other party. For several years adjudicators have generally accommodated any objection to alleged new argument or evidence by allowing the other party the opportunity to address it.

More recently, however, it was held that a decision was not enforceable because the adjudicator had considered a report not put to the other party during the discussion period prior to the adjudication. The report advanced arguments in respect of an extension of time claim that were different to those previously put forward by the referring party; different losses were also detailed. Here the court placed emphasis on a point frequently overlooked in the past, that the responding party must have been given the opportunity fully to consider the claim against them prior to adjudication – and that means all the supporting evidence. Moreover, whereas arguments may be refined or points that are no longer considered meritorious may be abandoned, a party cannot abandon wholesale arguments previously advanced and facts previously relied upon, and contend that the dispute is the same simply because the claim is the same[82].

Needless to say, in the light of the rather relaxed approach to such considerations previously adopted by the courts, this judgment has resulted in some confusion. Until further guidance is obtained it is suggested that referral notices should err on the safe side and not contain anything that the responding party has not already seen.

The Scheme lays down the minimum content of the referral notice. Standard form provisions and bespoke procedures state similar requirements as to what the referral notice should include; but even if yours does not you should not include less than the items listed by the Scheme, paragraph 7(2):

> ... *copies of, or relevant extracts from, the construction contract and such other documents as the referring party intends to rely upon.*

5.6.2   *Format of the notice*

Your referral does not have to follow any precise pattern and should be tailored to suit the nature and extent of the dispute. However, there is a general format that should be used as a guide.

5.6.2.1   Introduction

An introduction includes:

- The names and details of the parties
- The nature of the project and of the contract work
- Nature of the contract or service agreement
- The form of contract or agreement and date thereof
- Any incorporated documents that are relevant to the dispute
- The applicable adjudication procedure or rules (if not already incorporated into the contract or agreement)
- Details of the dispute or issues in dispute.

This list is not necessarily exhaustive but common sense will dictate what should be included in your particular case.

5.6.2.2   Narrative

A narrative states:

- The history of events giving rise to the matters in dispute
- Any alleged breaches of contract or agreement
- The consequences of any alleged breach
- Any other relevant information.

The narrative outlining the events and circumstances relating to your claim may be fairly lengthy if, for example, you are a contractor or subcontractor and are claiming disruption and delay, or you are a quantity surveyor, engineer or architect claiming additional fees from your client on account of additional work you have undertaken as a result of the manner of your client's instructions. On the other hand the narrative may be quite short if, for example, you are simply claiming non-payment of an undisputed sum.

   Your claim will be the consequence of failure of the other party to operate the terms of the contract properly and reasonably, or upon a breach of those terms or implied terms. You must state clearly the basis of your considered entitlement and what it is you are relying upon.

The success of your claim (unless it is a simple case of non–payment of monies due) will depend on your reasonably establishing a link between cause – the breach – and effect – the delay, additional costs incurred or suchlike. In legal language, you must establish a causal link. Remember that the burden of proof is on the person who asserts or alleges.

Always make your narrative comprehensive but not verbose and repetitive. The degree of success of a referral is not measured by the weight of paperwork submitted in the referral notice. Ensure, however, that it covers all issues in respect of which you are claiming. Do not include what are known as 'throw away' items that you have previously added to your claim submissions for good luck. That is a negotiating ploy and not appropriate to an adjudication submission. It will create a bad impression with the adjudicator and may adversely influence his decision as to who pays his costs.

If your claim involves quantum of any sort it should provide for the possibility that you will not get all that you are asking for. The claim might, for example, relate to additional cost in respect of which there are a number of contributory factors. If the adjudicator rejects some of the causes you will inevitably be entitled to only part of the sum claimed, but frequently referring parties make no request for any lesser amount as an alternative to the full amount. This should not create a difficulty but sometimes it does. To avoid problems you should not simply state something like:

'We claim £X'

but instead state something like:

'We claim £X or such other sum as the adjudicator may decide'.

The suggested wording overcomes the unwarranted concern of some adjudicators who feel that if they do not accept the full sum claimed they cannot award any other sum and must reject the claim in its entirety.

5.6.2.3   Redress sought

The redress sought includes:

- Consequential contractual entitlements or damages
- Questions to which answers are requested
- Declarations sought.

You should state clearly what it is that you are asking of the adjudicator.

It is not uncommon for an adjudicator to misunderstand what the referring party is requesting of him since a poorly presented referral may make it difficult for him to determine exactly what is being sought. There have been several instances where decisions have been set aside in whole or in part because it was held that the adjudicator, because he had misunderstood the referral, had decided matters that had not been referred to him. It is therefore good practice and helpful to the adjudicator, particularly if the dispute is complex, to conclude your narrative with what is termed the redress sought, i.e. a summary list of the items in respect of which you are seeking decisions and the amounts you are claiming. In so doing you will assist in defining the adjudicator's jurisdiction in the event of uncertainty arising[65].

It may be that you are merely seeking to establish certain principles, for example did the design brief adequately reflect your client's requirements, was work in accordance with the specification, were drawings issued late? In such cases it is common practice to pose questions which the adjudicator is requested to answer.

Redress sought should include a request that the adjudicator orders the other party to pay the whole of his costs (unless the adjudication procedure dictates otherwise). A similar request will need to be made in respect of your own costs if you have agreed to give the adjudicator power to allocate parties' costs.

5.6.2.4 Supporting documentation

Supporting documentation includes:

- Correspondence, etc.
- Records
- Legal judgments relied upon.

Your submission will need to include copies of all the documents necessary to substantiate your assertions. Thus, if you are a contractor or subcontractor alleging additional costs as a result of disruption and delay, these may include tender documents, programmes, correspondence, minutes of meetings, plant and labour returns, invoices, etc. If you are an engineer or architect claiming against your client on account of additional work you have had to undertake, you will need to provide details of the agreement with your client, his subsequent instructions, your staff daily work records and so on. Depending on the nature of the dispute you may need to include an expert's report provided the responding party has seen it or there is agreement to the inclusion of newly prepared reports. If you are relying on legal judgments these, or the relevant parts, should be provided.

To assist the adjudicator you should properly index and reference your documents so that if you make a point in your narrative the appropriate back-up document can be readily found. An adjudicator will get frustrated if he has to waste time wading through files of papers to locate a particular letter. Make sure that any drawings (particularly if you have reduced these in size) are legible and that any document that has been reduced in size can be read without a magnifying glass. Avoid black and white copies of coloured photographs since these are inevitably undecipherable. Photographs should be dated and a description given of what they are intended to demonstrate. Your supporting documents should be located in appendices which are properly referenced and subdivided if appropriate.

Remember when preparing your submission that you have to convince the adjudicator who is completely unfamiliar with your contract and the events that have taken place. If he has difficulty in comprehending your arguments or following any calculations, you will likely lose out because of the short time that he has in which to examine and understand the submissions and to make his decision.

As the referring party you have the benefit of time to prepare your submission and you should take advantage of this and ensure that you do not find yourself in the position of scrambling to get the referral notice completed by the due date.

Although the general method of communication when adjudicating is by fax, commonly without a follow-up hard copy, you should avoid sending the often large quantity of substantiating documents this way unless it is absolutely necessary. If you are up to the deadline you may have to send the files of substantiating documents by courier although there should be no problem if the adjudicator and the other party get these the day following the referral notice proper.

5.6.2.5   Summary

The referral notice is your opportunity properly to present your case. It is most undesirable to have to make amendments or corrections later. Apart from making yourself unpopular with the adjudicator, to do so may undermine your credibility.
Ensure that:

(1)   The dispute you are referring has not already been referred and decided in whole or in part
(2)   Your notice includes sufficient of the contract documents to enable the adjudicator to satisfy himself that you have a valid contract, that it is a construction contract and that it is caught by the Act
(3)   Your notice is complete, it addresses all issues in respect of which you are claiming and clearly identifies what it is that you are asking the adjudicator to decide
(4)   Each issue is supported by appropriate substantiation
(5)   All copy documents are legible
(6)   All documents are indexed and referenced
(7)   You have given the adjudicator your contact telephone and fax numbers, i.e. those of the person who will be handling the adjudication on your side
(8)   Your notice is sent on time
(9)   A copy of the notice is sent simultaneously to the responding party.

5.7   What if you are unable to submit the referral notice on time?

The Act at section 108(2)(b) states that any adjudication procedure must provide a timetable with the object of securing both appointment of the adjudicator and referral to him within seven days of the notice of adjudication. Is absolute adherence to the seven days necessary or is there some latitude? Is a referral notice still valid if it is served later than seven days? There is some difference of opinion.
The use of the words *with the object of* suggests perhaps that failure to meet the deadline may not be critical. However, the Scheme for example is more specific. Paragraph 7(1) states:

> ... the referring party shall, not later than seven days from the date of the notice of adjudication, refer the dispute ... to the adjudicator ...

Other procedures are similarly specific and, in law, *shall* means *must.*

However, the time-scales set down in the Act are there primarily for the benefit of the referring party by ensuring that the ANB nominates promptly and the adjudicator does not delay in responding to the referring party's request for a decision. It may be argued therefore that lateness on your part should not make any difference provided of course that the other party is not prejudiced or you are not behaving unreasonably. You may have been prevented from keeping within the seven days because the ANB for some reason did not make an appointment in time; indeed the ICE Procedure contemplates the possibility that an appointment itself, let alone service of a referral notice, may not be achieved within seven days.

As soon as you are aware that you will exceed the seven days, advise the adjudicator explaining why your referral will be late and request such further time as you require. The other party may be awkward and argue that a late referral is invalid but it will be up to the adjudicator to decide whether to continue or not; at worst you will have to start again. If he does continue, it will be of comfort to know that among the published judgments there is no case where enforcement proceedings have been unsuccessful because the referral notice was late.

6.   The conduct of the adjudication

6.1   The process after referral

Whilst conforming to a general pattern, the conduct of adjudications varies considerably. There are no set rules and the actual procedure will depend to a large extent on the nature and complexity of the dispute. All is in the hands of the adjudicator who has wide powers to enable him to conduct the adjudication in the most effective manner, although its progress will be influenced by the aggressiveness and determination of the parties in pushing or obstructing the process. The adjudicator's powers are derived from the Act or from the applicable adjudication procedure, or they may be implied, i.e. as if the Act or procedure had expressly specified them.

The Act in fact says very little on the adjudication process. It empowers an adjudicator to take the initiative in ascertaining the facts and the law (section 108(2)(f)). This means, for example, that he may make enquiries of third parties in order to obtain information to assist him in reaching his decision, although he must advise the parties to the dispute of anything he finds out. Otherwise the Act itself does not say anything regarding the adjudicator's powers, but it does exercise some control in that it requires him to act impartially (section 108(2)(e)) and to produce his decision on time (section 108(2)(c)).

Without a more specific indication as to what precisely an adjudicator may do in the exercise of his duties, serious difficulties could arise through challenges to his behaviour by one or other of the parties. Such challenges would inevitably fail provided that, in the context of the nature of adjudication, the actions of the adjudicator were reasonable. However, much time and cost could be wasted. Hence rules and procedures spell out some of the more common powers that are necessary and which would otherwise be implied to enable the proper conduct of an adjudication.

Under the Scheme the powers given to the adjudicator to give instructions, or directions as they are commonly called, to the parties are set down at paragraph 13:

(a) request any party to the contract to supply him with such documents as he may reasonably require including, if he so directs, any written statement from any party to the contract supporting or supplementing the referral notice and any other documents given under paragraph 7(2).

(b) decide the language or languages to be used in the adjudication and whether a translation of any document is to be provided and if so by whom.

(c) meet and question any of the parties to the contract and their representatives.

(d) subject to obtaining any necessary consent from a third party or parties, make such site visits and inspections as he considers appropriate, whether accompanied by the parties or not.

(e) subject to obtaining any necessary consent from a third party or parties, carry out any tests or experiments.

(f) obtain and consider such representations and submissions as he requires and, provided he has notified the parties of his intention, appoint experts, assessors or legal advisers.

(g) give directions as to the timetable for the adjudication, any deadlines, or limits as to the length of written documents or oral representations to be complied with, and

(h) issue other directions relating to the conduct of the adjudication.

Failure to comply with a reasonable direction will be to the detriment of the defaulting party.

The Scheme at paragraph 12 places certain restraints on the adjudicator in addition to the two controls stated in the Act. He is required to carry out his duties in accordance with any relevant terms of the contract, to reach his decision in accordance with the applicable law, and he must avoid incurring unnecessary expense. In fact, apart from the period for reaching his decision, all the controls and restraints set down in the Act and the Scheme would probably be implied anyway but it is helpful to have them spelled out.

Other standard procedures contain provisions giving similar powers to and exercising similar controls over the adjudicator. But even if a particular procedure does not do so, the sort of powers and controls set down in the Scheme would be implied by law. It must be remembered that adjudication has always been accepted as being a somewhat rough and ready process. The requirement that a decision, perhaps on a complicated dispute and often based on limited information, is to be given – in the context of legal processes – in a very short time means that there will inevitably be some rough justice along the way. The niceties of arbitration or litigation cannot always be observed and the courts, when investigating adjudicators' behaviour, have acknowledged this. However it is essential that the adjudicator acts fairly and conducts the proceedings in an even-handed manner. (See Part 4, section 1.3, in this book.)

Having received your referral notice the adjudicator, if he has not already set down a timetable, will normally invite the responding party to submit their defence, commonly called a response, to your claim, stating the period within which this is to be done – although the procedure itself may state a time; the JCT Procedure for example allows seven days after referral.

6.1.1   The responding party's response to the referral notice

The response should be confined to addressing the issues, evidence and argument put forward in the referral notice. You should check it carefully to see if it goes further. The introduction into submissions of new material is discussed at Part 4, section 1.9, in this book.

6.1.2   Your reply to the response

You should be invited by the adjudicator to respond to the response but if you are not, it is good policy to do so anyway. Your reply, as it

may be called, should be restricted to any matters raised in the response which you did not address in your referral notice. However, notwithstanding any restrictions that the adjudicator may have placed on the scope of your reply, you should not allow to go unchallenged any allegation or assertion in the defence with which you disagree.

You must not introduce new aspects unless they are material to challenging what the defence has stated. You may possibly wish to submit further supportive documents but you should be cautious as regards introducing anything that you have not already put to the responding party (see Part 4, section 1.9, in this book). Whatever you do, do not repeat what you have already stated in your referral; this will irritate the adjudicator and give him more work in sorting duplications. Remember he needs as precise but as concise a picture as you can give him.

6.1.3   Further submissions

It is not uncommon for the parties to continue to make submissions right up to the time the adjudicator is due to make his decision. He has to allow himself time to write his decision, usually not more than one week or less than four days, and the wise adjudicator will state a cut-off date after which he will not accept submissions. Even so there is no harm in sending something to him which you consider is necessary for the justification or support of your case.

You will need to react promptly to any directions of the adjudicator such as a request for further information or clarification of your referral notice or any subsequent submission you have made. Indeed you must deal with anything requiring your attention without delay. You had all the time you needed in which to prepare your referral notice but with subsequent submissions time is as much against you as it is against the responding party.

There may be a need for a site inspection and it is common for adjudicators to convene a meeting of the parties; these matters are entirely at the discretion of the adjudicator. (See Part 4, section 1.4 on meetings and hearings and section 1.5 on site inspections.)

6.2   Refusal of the responding party to participate

If they did not do so at the time of the adjudication or referral notices, the responding party in their defence may question your right to refer your claim at all because, for example, they are of the opinion that the subject work is not caught by the Act or the adjudicator has been incorrectly appointed. Such a challenge may be expressed as a challenge to the adjudicator's jurisdiction. In that event the adjudicator should invite your comments. Hopefully you will have taken the precaution of thoroughly checking that there was no impediment to your referring your dispute to adjudication and you will have an answer ready. If however you have been caught off-guard, you need to take rapid action which may involve seeking advice from an expert or lawyer if you have been representing yourself.

An adjudicator is not allowed to decide, in the legal sense, his jurisdiction unless the operative adjudication procedure specifically permits this (of the standard procedures only the TeCSA rules do) or the parties agree that he can. But he must make up his mind whether to proceed or not. If he advises that he is not going to proceed, that is the end of the matter unless you raise a legal challenge to his conclusion. The court will then decide whether or not he was correct not to proceed. If, upon advice, you are convinced the adjudicator is wrong, you could as an alternative seek the appointment of another adjudicator who may take a different view as to jurisdiction. As your dispute has not been decided you are entitled to do this.

On the other hand, if the adjudicator advises that he is going to proceed he will probably require from you an undertaking that you will pay the whole of his costs in the event that a court may decide that he was wrong to do so.

If the adjudicator does proceed in the face of a challenge from the responding party, the latter is entitled to apply for an injunction to stop the adjudication. You will then have the option either to back down and not pursue your claim or to take up the challenge, which is going to incur you in cost most of which you should recover if you defeat the challenge.

Alternatively the responding party, having registered a challenge, may do nothing more and then, if the decision is in your favour, may either apply to the court for it to be set aside or simply refuse to carry it out and wait for you to do something. In that event you will have to decide whether or not to take enforcement proceedings.

6.3 Can you withdraw the referral notice?

As the referring party you may at any time stop the proceedings. Another commentary states that the agreement of the responding party is required but, it is suggested, that cannot be correct. Perhaps, having seen the defence, you decide you do not have such a strong case as you originally thought; or the responding party may have become bankrupt. It would be iniquitous if in such circumstances a referring party was forced to continue unless the responding party consented to termination of the proceedings.

If you do decide not to proceed you will have to pay all the costs that the adjudicator has already incurred, but you are not required to pay any of the responding party's costs unless you have agreed to a procedure which so states.

6.4   Can the responding party submit a counterclaim?

The adjudicator can deal only with the dispute referred to him. Consequently he has no jurisdiction to consider any counterclaim that has not been the subject of a valid withholding notice (see Part 5 of this book) unless you and the responding party agree to extend his jurisdiction (see Part 3, section 3.6, in this book).

However, the existence of a counterclaim may be relevant if you have to take enforcement proceedings (See Part 2, section 8.3, in this book).

7.   *The decision*

7.1   Format of the decision

The written decision is the adjudication equivalent of an arbitrator's award or a court judgment. It gives effect to the contractual agreement between the parties that disputes may be resolved, temporarily at least, by adjudication and enables the parties to be certain as to what the adjudicator has decided and, if necessary, enables enforcement through the courts. However, decisions are not always as clear as they might be[32].

The format will depend on the style of the adjudicator but a decision should have an introduction which states the details of the parties, the contract or agreement between them, the manner and date of the adjudicator's appointment and the dates of the notice of adjudication and the referral notice. The dispute should be identified and it is good practice for an adjudicator to list what it is he considers he is being asked to decide if this is not absolutely clear from the referral notice.

It is not necessary for the content of a decision to be as extensive as an award or judgment where contract clauses, correspondence and the like are reproduced often verbatim and evidence discussed in detail. In fact unless reasons are requested there is no requirement for the adjudicator to include anything beyond the introduction other than a statement as to what he has decided. But in practice a decision will normally include some background information to the dispute and a summary review of the more important evidence indicating, where there is conflict, which version the adjudicator has accepted. He will then state the conclusions he has reached.

Finally, the adjudicator will state his decision, which will probably comprise a number of decisions – one for each issue. If something is required to be done then a period for compliance should be stated, although 'immediately' and 'forthwith' are sometimes used. If a sum of money is to be paid it should be made clear whether or not value added tax is included. Normally monetary awards are exclusive of value added tax but it has been held that an adjudicator may include in his decision the amount of tax payable, if requested by the notice of adjudication[127].

A decision is not complete without a statement of the adjudicator's costs and how they are allocated between the parties. If the adjudicator is empowered to award party costs then the decision must state how these are to be dealt with.

The decision must be signed and dated, but it is not necessary for the adjudicator's signature to be witnessed, although many are.

7.2   A reasoned decision

A reasoned decision is one which includes a comprehensive explanation of the adjudicator's thinking to enable the parties to understand how he has come to that decision. This may require a lengthy consideration of conflicting evidence and arguments. Where given, reasons may be included within the narrative of the decision or they may be set down separately – even as an attachment. Reasons form part of the decision and how they are presented makes no difference to the significance of what they say.

Whether an adjudicator can be required to provide reasons will depend on which procedural rules govern the adjudication. The entitlement of the parties to require a reasoned decision is considered at Part 4, section 2.2, in this book.

7.3   Obtaining the decision

The adjudicator will advise the parties when his decision is ready and he will indicate the release procedure. There will be a copy for each party and it will probably be possible for these to be collected from the adjudicator's office. Otherwise they will be posted, in which case it should be by some express form of delivery. Adjudicators will often agree to send the decision initially by fax or e-mail.

From time to time adjudicators experience difficulty in recovering their costs from the parties. Consequently many endeavour to protect themselves by claiming a lien on their decision until payment is received.

In their terms and conditions adjudicators may well therefore include a stipulation to the effect that the decision will not be released until payment of their costs has been made in full. If the adjudicator has asked for prepayment and you have not clearly rejected that request, you may be held to have accepted it and if you want to get your hands on the decision then there will be no option but to pay in advance. But who pays at this stage?

If the adjudicator has exercised a lien on his decision it is a good idea at commencement of the adjudication to try to obtain the agreement of the other party that each pays one half of the adjudicator's costs in order to secure the decision. Otherwise, as the party who started the process, you will have to pay all of his costs in the first instance. In either situation a reconciliation is carried out when the adjudicator's decision as to apportionment of his costs is known.

If he has not already obtained the parties' agreement through his terms and conditions then the adjudicator may in an early direction, or when his decision is ready, state that he will require prepayment. Opinion differs as to whether, in the absence of prior agreement of the parties, he can do this. In the early days of statutory adjudication general opinion was that he could; currently opinion has swayed the other way. The Royal Institution of Chartered Surveyors recommends to its adjudicators that they should not attempt to demand prepayment; the ICE Adjudication Procedure states that an adjudicator may do so; the TeCSA Rules preclude an adjudicator from requiring any advance payment for his fees; the Scheme is silent on the matter. So you need to check the relevant procedure before taking issue.

If the adjudicator has not established an entitlement to prepayment but refuses to send the parties his decision until his costs are paid, your only recourse, if you do not want to pay in advance, is to seek a court order directing that he releases his decision. On the basis that the adjudicator is entitled to reasonable payment anyway, the court, whilst not approving of the adjudicator's behaviour, might be reluctant to give such an order.

But ask yourself why it is that you do not wish to pay in advance. Is it that you want to see the decision first? If so, why? Do you intend not to pay the adjudicator's costs if you consider the decision to be wrong in some way? If so, think again. Even though the adjudicator may have come to the wrong decision, this does not mean he is not entitled to his reasonable costs.

Only if you intend to challenge the decision on the grounds that the adjudicator misconducted himself to the extent that the decision should be set aside, should you refuse to pay his otherwise reasonable costs. If the court agrees with you and considers that the adjudicator acted in bad faith, it will likely order that he is not entitled to payment.

7.4 What if the adjudicator has not dealt with all the issues referred to him?

Failure on the part of the adjudicator to address all matters referred to him does not invalidate the decision; such an omission on the part of an adjudicator has been held to be an error and the decision is still valid and enforceable[18/41]. The court will not intervene to address the omission. However, an adjudicator who decided he could not deal with a dispute despite having jurisdiction was held to be wrong by both the Court of Session[10] and the Appeal Court[124].

An omission may be considered as a slip capable of being corrected and you should put this to the adjudicator. If you can obtain the support of the other party so much the better since the adjudicator will then know that any action he takes to rectify the omission will not be challenged.

Hopefully he will address your complaint by way of an amendment to his decision but he must only address the omission. He cannot alter anything else he has done unless he admits that he has made other slips. If the adjudicator declines to deal with your complaint then you have two options:

(1) If he declined to deal with the subject matter of the omission because he considered he did not have jurisdiction, try to obtain the agreement of the responding party to extend his jurisdiction. It is in order for an adjudicator to deal separately with different issues at different times.

(2)   Commence a second adjudication on the omitted issue(s) hopefully with the agreement of the responding party that the same adjudicator should act, thereby minimising costs.

7.5   What if you consider the adjudicator's costs to be excessive?

There is one good reason why you may not want to pay the adjudicator's costs: you consider them to be unreasonable. Even though you may have in some way accepted his hourly rate, you have not agreed to the number of hours he may charge and perhaps you consider these to be excessive in the context of the work he has done. Overcharging in this way is the basis of regular complaints against adjudicators. However, an adjudicator is entitled to reasonable payment for his services even if for some reason he gets things wrong and his decision is set aside or returned to him by the court for adjustment or correction.

If you do consider the adjudicator has overcharged, how you deal with it will depend on the circumstances.

If you are not obliged to prepay for the decision, even though you would otherwise have made payment voluntarily before receiving it, you can, when you find the amount he is charging to be excessive, call upon the adjudicator to release the decision telling him that you intend to challenge the amount of his costs.

If the adjudicator has established a lien and you are obliged to prepay for the decision, you have a dilemma. The object of adjudication is to obtain a quick decision. The last thing you want is a long delay while the amount of the adjudicator's costs is sorted out, so probably your best course of action is to register a protest, pay up to secure the decision and then argue about the amount later. The challenging of adjudicators' costs is dealt with at Part 4, section 2.4, in this book.

8.   *Enforcement of the decision*

8.1   What if the responding party does not do as the adjudicator has ordered?

Sometimes the referring party is simply seeking a decision on a point of principle, the interpretation of some aspect of the contract, the validity of an instruction, or the like. Rarely does an adjudicator direct a party to carry out a physical activity – what is known as ordering specific performance. Failure to comply with a decision normally amounts to a refusal to pay money. There are various reasons for such obstructive behaviour on the part of the responding party:

(1) They do not realise what adjudication is all about and that the decision is enforceable
(2) They are just being awkward and hoping you will not pursue your entitlements
(3) They consider that the adjudicator did not have jurisdiction – either at all or in respect of perhaps certain aspects of the decision
(4) They consider that the adjudicator has not conducted the proceedings properly or he has been biased and this has led to an unfair decision
(5) They consider the decision is in error
(6) They are not in a position to pay what the adjudicator has ordered
(7) They consider that they have a claim which can be set off against the decision.

If you suspect that the reason why the responding party is not complying is (1) or (2) above, then an appropriate letter from your solicitor may produce a result. If not, and if non-payment is for any of the other reasons, then enforcement proceedings in the courts must be contemplated. Before taking this step it is essential that you obtain advice from someone experienced in adjudication and familiar with the manner in which the courts have dealt with challenges to enforcement.

The enforcement process is much simpler than was first thought at the time statutory adjudication was introduced. Then it was considered that the successful party would need to obtain an arbitrator's award stating that the adjudicator's decision must be complied with. If the unsuccessful party still refused, then summary judgment – a legal process available where there is no viable defence – could be obtained. The prospect of having to go through that long process was somewhat daunting.

Fortunately on the first occasion that the matter was put to the test, the arbitration step was omitted and the referring party applied directly for summary judgment, requiring the responding party to honour the decision. The court upheld the application and the decision was enforced quickly and simply[66]. That case set a precedent for all future applications for enforcement. An alternative procedure is to serve a statutory demand on the responding party.

The comparable procedure in Scotland is judicial review, in which the applicant party petitions the court. Judgments and court transcripts are normally entitled Petition of X rather than X v Y as in England and Wales, although commentators often introduce the defendant's name when referring to Scottish cases. Judicial review is not available in England and Wales except in respect of issues of public law.

Generally, in contrast to the normal pace of legal processes, the courts have when asked responded to the spirit of the Act; applications to the courts relating to adjudication matters have been dealt with promptly. Indeed, a delay in complying with a decision can give rise to difficulties as regards costs if proceedings are commenced before payment is made[84].

A special contractual clause restricting when legal proceedings may commence was held to be ineffective as regards enforcement of an adjudicator's decision[121]. And where appropriate there is no bar to other aspects of the legal process, such as a winding up petition being included[93]. If yours is a Scottish contract your summons may include the traditional warrant for inhibition (a means of obtaining security for the amount involved) but, as a result of a judgment in 2002, this is not automatically granted as it used to be[111].

In all but the most exceptional cases, a court merely requires to be satisfied that the adjudicator had the requisite jurisdiction and had conducted the proceedings in accordance with the so-called rules of natural justice or, using contemporary terminology, with procedural fairness.

Consequently, other than a spate of earlier cases concerning alleged non-compliance with payment procedures, reasons (3) and (4) in the above list provide the majority of applications for enforcement of adjudicator's decisions. Reason (3) is discussed at Part 3, section 5.1 and reason (4), being a matter of common interest, at Part 4, section 3.1, in this book.

As regards reason (5) in the above list, the merits of the decision will not be investigated and it will not be set aside merely because it contains errors (see Part 4, section 2.3, in this book).

The more difficult situations arise if the basis for non-compliance with the decision is centred on reasons (6) or (7) in the above list.

If your methods of more gentle persuasion have failed to convince the responding party to pay up as ordered by the adjudicator, it is essential that you obtain good advice before embarking on enforcement proceedings. As with most legal matters the outcome can never be guaranteed. Many of the judgments given by the courts are contradictory and commentators on adjudication matters acknowledge that the position regarding enforcement is confusing where there are considerations of set-off, counterclaim, other current proceedings or insolvency. A perusal of a few of the cases relating to enforcement will demonstrate the point.

However, do not waste time on enforcement proceedings if the monies that the adjudicator has ordered the responding party to pay you have been obtained fraudulently. Your application to the court will not succeed[89].

8.2 What if the responding party contends that they are unable to pay you?

You may not have been aware that the party against whom you are claiming was in financial difficulties. Even if you knew, perhaps the situation has deteriorated since you commenced the adjudication and, worse, the responding party has gone into liquidation, in which event there is probably no point in pursuing your entitlement any further.

But the fact that the responding party has financial problems is not good reason in itself for an adjudicator's decision not being enforced. Unless the responding party is in financial difficulty to the extent that they can argue that payment to you of the sum ordered by the adjudicator will force them into insolvency, a court will normally grant summary judgment[88]. It has been stated in the course of enforcement proceedings that, if appropriate, the courts will look at the financial circumstances of both parties[54].

Adjudication has been held to be a legal process for the purposes of the Insolvency Act 1986[108], and if the responding party is in liquidation the Act prevents any action being taken against them without the permission of the courts. The same applies if they are in administrative receivership, whilst an administration order remains in place. Because administrative receivership does not have the finality of liquidation, the attitude of the courts when considering enforcement of a decision will be influenced accordingly.

It is the combination of the insolvency of the referring party together with the existence of a counterclaim of the responding party or active or pending recovery action on their part, that has provided the background to most successful challenges to enforcement of valid adjudicators' decisions.

8.3 What if the responding party refuses to pay on the grounds of a set-off or counterclaim?

A number of judgments have dealt with situations where the losing party has argued against paying the amount of the adjudicator's decision on the grounds that this party had a set-off or counterclaim. Generally such arguments have been unsuccessful.

If the responding party considered it had a justifiable set-off against your claim, or some sort of counterclaim, this should have been the subject of a proper withholding notice prior to the adjudication. The chances are therefore remote of it successfully defending enforcement proceedings on the basis of some considered entitlement against you[1/4/70/79].

The position is not affected if the contract between the parties contains wording permitting set-off, although there has been some confusion in reaching this conclusion. An initial Court of Appeal decision (albeit in respect of a non-statutory adjudication) held – apparently with some difficulty – that specific set-off provisions in the contract prevailed and could be utilised if an unsuccessful responding party wished to set off against an adjudicator's decision[87].

Shortly after in a statutory adjudication case it was held that specific set-off clauses were not effective[64]. This decision was appealed but the Court of Appeal upheld the lower court[132], although in the interim period another judgment had found to the contrary[17].

There has been some contradiction on the matter of withholding notices given in respect of an adjudicator's decision. In an early judgment it was said that to be able to withhold against a decision would make a nonsense of the principles of adjudication, and the decision was enforced[113]; a similar view was taken in a later case[31]. A notice in respect of the withholding of liquidated damages given before the due date for payment of the amount of a decision was held to be ineffective[81]. But in another case a withholding notice given in time after a decision, was held possibly to be valid[71] and summary judgment was refused.

Where a responding party had commenced arbitration claiming an entitlement greater than the amount of the adjudicator's decision, the decision was still enforced against it even though the responding party argued that the referring party would not be able to pay if it (the responding party) were successful in the arbitration. Here, it would appear, the court was not sufficiently concerned about the financial state of the referring party to refuse summary judgment or grant a stay. In later similar situations the courts have tended to react differently.

An alleged previous overpayment to the successful referring party was held to be no bar to enforcement of the decision since the matter of overpayment was outside the scope of the adjudication[117].

It has also been held that the existence of a dispute on another contract between the parties does not affect enforcement[104]. This followed an earlier case where it was held that an adjudicator cannot order that payment of a sum be deferred pending resolution of other disputes[5].

However, the situation is different where the responding party may be forced into bankruptcy if it had to pay the amount of the adjudicator's decision, but has a valid counterclaim in respect of which there are ongoing proceedings. In such cases the courts have to take account of the insolvency rules (the Insolvency Act 1986) and will probably refuse enforcement[88]. This topic is considered at Part 3, 5.3, in this book.

8.4 *Would enforcement proceedings be successful if you became insolvent?*

If on the other hand it is you, the successful referring party, who becomes insolvent, then it will be the receiver or administrator who through you will apply for summary judgment of the adjudicator's decision. In that circumstance a challenge to enforcement by the responding party on the basis of a considered counterclaim of some sort is far more likely to be successful[16].

Then, the responding party may argue that if it pays the amount of the adjudicator's decision, it will be unable to recover the amount (or even part of it) at a successful subsequent arbitration[123]. Or the responding party may contend that it has a counterclaim which is equal to or greater than the amount of the decision and which it is unable to secure because of your insolvency; why therefore should it pay you anything? The insolvency rules dictate how the situation is dealt with.

The onus is on the referring party to demonstrate that it would be able to pay any future successful claim against it before summary judgment is given[92]. Clearly to do so would be somewhat difficult in the case of insolvency but may be possible when the referring party is in administrative receivership.

Defences to enforcement proceedings are also discussed at Part 3, section 5, in this book.

Part 3
So You Are Being Taken To Adjudication

PART 3

SO YOU ARE BEING TAKEN TO ADJUDICATION

1. Initial steps

1.1 What do you do when you find you may be taken to adjudication?

The other party to your contract may threaten to refer to adjudication a dispute about which you have been arguing. They may be bluffing and hoping to frighten you into conceding. If you think you are in the right then stand firm and do nothing but if you have simply been stalling, now is the time to be realistic. You will not want to expend effort and incur the costs of an adjudication unnecessarily, so try and reach a settlement unless the other party is pursuing something to which you genuinely consider they are not entitled.

1.2 Can you refuse to participate in an adjudication?

There is no sanction if you refuse to participate in a valid adjudication, but should you choose not to do so, then in the absence of a defence from you, the adjudicator will invariably decide wholly or largely in favour of the referring party simply because he has no reason not to. If however the referral is defective in some way or the adjudicator lacks jurisdiction, you may be able to halt the process temporarily or permanently or, if it proceeds, get the decision set aside in due course.

The criteria for a valid referral have been discussed at Part 2. It follows that the same criteria are relevant if you are seeking a means of avoiding, or at least deferring, participation in the proceedings.

Possible grounds for challenging the validity of a referral are:

- You have no contract with the referring party
- Your contract was not in writing as defined at section 107 of the Act
- Your contract is not a construction contract as defined at section

104 of the Act, or by virtue of section 105(2), it is not a construction operation

- Your contract is excluded under sections 106(1) or 117(1) of the Act or by the Construction Contracts (England and Wales) Exclusion Order 1998
- Your contract was entered into before 1 May 1998
- There is no dispute in respect of the matter or matters being referred
- The dispute being referred has already been the subject of adjudication
- If the Scheme applies, then rather than arising under your contract (paragraph 1(1)), the dispute is in connection with or related to, or it has some similar more distant relationship
- There has been a settlement

and, subject to the provisions of the adjudication rules or procedure:

- More than one dispute is being referred
- The adjudicator is already conducting a related dispute under another contract
- The adjudicator has been incorrectly nominated.

1.3   What do you do when you receive a notice of adjudication?

If a notice of adjudication is served on you it must never be ignored whatever your feelings as to the validity of the claims. The fact that you would rather not get involved in an adjudication is of no consequence; the other party's action probably has serious intent, in which case you have no choice. You refuse to participate at your peril unless there are circumstances which undeniably preclude the referring party's right to proceed.

If the notice alone is received and you accept that it is valid there is nothing to do at this point.

However, unless an adjudicator is named in the contract, the notice may be accompanied by the name of a proposed adjudicator or perhaps several names from which you are invited to make a selection. If you are offered a name whom you consider (from your knowledge of the individual) would be suitable, then you should accept him; otherwise the referring party will request an ANB for a nomination over which you will have no control. When you are considering a proposed name remember that anyone you accept needs to be experienced in adjudication as well as being technically competent. And you should not accept him if there is some connection or association, either personally or professionally, between you that might give rise to a suspicion of bias. Otherwise when this comes to light the appointment may be invalidated.

Alternatively, the referring party may not be interested in trying to agree an adjudicator and you may, with the notice of adjudication, receive a copy of their request to an adjudicator nominating body. In that case check the application to see it is correct:

- Is information given about the nature of the dispute, the contract and the parties, accurate?
- Does it correctly state the expertise that the adjudicator should have?
- Was it sent to the correct ANB?

It is important that the information given to the ANB is accurate to ensure, as far as possible, that a suitable adjudicator is nominated. If there is time (and remember that the ANB probably has only five days in which to make a nomination) you may contact them stating your own views should you consider that the referring party has given wrong or misleading information. The ANB should take account of your comments and should do all it can to accommodate the wishes of the parties, but it has no obligation in this respect. If you do write to the ANB ensure that you copy the letter to the referring party.

If you do not have time to express your views beforehand, you should take up any concerns that arise from the appointment immediately. What you may then do is considered in the next section.

2. Objecting to becoming involved

2.1 What do you do if you consider you have a valid objection to the adjudication proceeding?

If you consider that the intended referral is invalid you should draw your objection to the attention of the referring party at the earliest opportunity, preferably on receipt of the notice of adjudication. If you miss this opportunity or the referring party does not accept your objection, then as soon as the adjudicator is in place write to him setting down your reasons as to why you consider the referral is invalid. Do not waste time and effort making your objection to the adjudication nominating body. The policy of such bodies is that they cannot become involved in jurisdiction arguments and they leave it to the adjudicator they have appointed, to decide whether he should proceed or not. The only exception may be if you are contending that the adjudicator was incorrectly appointed.

In raising an objection you are alleging that the adjudicator does not have jurisdiction either generally, because for example the contract is not caught by the Act, or specifically because of the particular circumstances, for example that there is no referable dispute.

Unless the adjudication procedure or rules permit, an adjudicator cannot decide his own jurisdiction[41/90]. If the operative procedure allows, you may challenge his jurisdiction and ask him to decide it. If the procedure does not allow this, you may be able to persuade the referring party to agree to widening the adjudicator's powers to include the dispute over jurisdiction[78]. The adjudicator will invite submissions from both parties and will reach a conclusion as to his jurisdiction.

It is however most important to appreciate the significance of the adjudicator having the power to decide his own jurisdiction – whether this comes from the adjudication procedure or through the parties having given it to him.

If the adjudicator decides that he has jurisdiction but you disagree, you cannot use that disagreement as a defence in any subsequent enforcement proceedings against you. His decision on jurisdiction is a decision by which the parties have agreed to be bound[78/117]. So if his decision is in favour of the referring party, you have no option but to get on and defend the claim against you. It should be mentioned that an earlier Scottish judgment reached the opposite conclusion but the particular facts were influential[56].

If the procedure does not permit or the adjudicator is not given the power by the parties to decide his jurisdiction, then he has to make up his mind whether to proceed. Effectively he has to conclude whether or not he has jurisdiction[28]. Without the power he cannot decide this – using the word decide in its legal sense. Instead he has to consider the facts and arguments, again by way of submissions from the parties, and then advise whether or not he intends to proceed. His conclusion – whatever it is – can be reviewed by the courts as it is not an adjudication decision.

It is helpful if the adjudicator gives his reasons for arriving at his conclusion, but he is not obliged to do so. If he decides he does not have jurisdiction and resigns, the referring party may apply for the appointment of another adjudicator. The ANB approached will not question the situation and will make an appointment. If the second adjudicator correctly concludes he does have jurisdiction his decision is valid. The referring party has a right to have its dispute decided[114].

If the adjudicator states that he is proceeding and gives his reasons, these may persuade you that your objection has no validity. If he offers no explanation or you disagree with it, then you must decide between the following three courses of action.

2.1.1 Stand aside, let him and the referring party proceed without you and then refuse to carry out the decision if it goes against you

A decision not to participate should be taken only if you are absolutely sure that you are right. Otherwise the adjudicator will make his decision based solely on the evidence in the referral notice and may well find totally in favour of the referring party. If later you do not execute the requirements of the decision then enforcement proceedings will probably be taken. Should the court decide you were in the wrong and that the adjudicator did indeed have jurisdiction, it may end up costing you more than would have been the case had you defended yourself. If the dispute is a complex one you may of course decide that the cost of conducting a defence would be so great as to warrant that risk.

2.1.2 Participate without prejudice to your view that you are not required to do so, and then refuse to carry out the decision if it goes against you

If you decide to participate, it is essential that you record with the adjudicator and the referring party that you are doing so without prejudice to your opinion that he lacks jurisdiction and your right subsequently to challenge it. Failure to do so will almost certainly result in your subsequently being held to have waived your right to object and you will not be permitted to raise a challenge[69]. Some objecting responding parties endorse all communications 'without prejudice'. That may be over-cautious but better safe than sorry.

You should take care not to debate the point of jurisdiction in your response or in correspondence. In so doing you may inadvertently confer authority on the adjudicator to decide his jurisdiction[77].

Provided you have registered your objection your course of action, if the decision goes against you, would be to refuse to honour it, giving your reasons, and then use your objection as your defence when the referring party takes enforcement proceedings.

2.1.3 *Seek the intervention of the court*

The uncertainty associated with having to wait until after the decision before the position is established, can be avoided by an immediate application to the court either for an injunction restraining the adjudicator or for a declaration that the adjudicator lacks jurisdiction. Such an application would be heard very swiftly. It would be beneficial if the referring party would agree to extend the time in which the adjudicator has to make his decision – effectively to put the adjudication on hold – until the court has given its ruling. But should the referring party not agree, and in all probability it would not, the court's ruling could still be to hand before the adjudicator's decision is due.

Seeking a declaration from the court is probably the best course of action if the dispute is complex and would involve substantial effort and cost to defend. Although an application is a straightforward process in itself, experience is that the courts prefer not to interfere with the adjudication process. That said there have been more than a dozen reported instances where they have been involved at the early stages of an adjudication.

2.2   What if you have reservations about the nominated adjudicator?

When you know who the adjudicator is you may be concerned for any of the same reasons as the referring party:

(1)   He is not suitably technically qualified
(2)   You consider him not to be appropriately experienced in adjudication
(3)   You have good reason to believe that he may not be impartial.

These will now be dealt with in turn.

2.2.1 *Technical qualifications*

The referring party ought to have provided the ANB with the maximum information to enable it to make a suitable appointment. However, referring parties do not always behave as they should and ANBs can occasionally be careless or make mistakes.

If the referring party sends you a copy of its application to the ANB (and it is not obliged to do so) and you see that it has requested the appointment of someone with expertise that you consider not to be appropriate, you may advise the ANB of your opinion, giving your reasons, and request that they appoint a person of more suitable experience than the referring party has suggested. Time is very limited at this point but it is your only opportunity to influence the ANB. The ANB does not have to respond and is more likely to stay with the advice of the referring party, but it is worth a try.

More probably the adjudicator will have been nominated before you can make your feelings known and, even though you may consider the selection inappropriate, there is effectively nothing you can do unless you can secure the co-operation of the referring party.

2.2.2 Adjudication experience

Here you will almost certainly be unable to do anything. Those selected by ANBs for nomination as adjudicators should have gone through a training and assessment programme to verify as far as possible their capability to adjudicate properly. Even if the requirements of a particular ANB are lower than they ought to be, the ANB will not want to admit that it may have inferior adjudicators on its list and almost certainly will not reconsider its nomination. The ANB would in any case be concerned that any attempt to do so might result in legal action against it by the adjudicator concerned.

Concerns regarding the competence of the adjudicator are discussed at Part 4, section 1.2, in this book.

2.2.3 Lack of impartiality

Anyone active in the fields of dispute resolution encounters a wide range of individuals, companies and organisations. It is possible therefore that an appointed adjudicator may have had some previous association or connection with one of the parties.

It would depend on the nature of such contact, its extent and duration and how long ago it occurred, as to whether it might be considered as rendering the adjudicator vulnerable to bias or lack of impartiality.

The adjudicator himself should not accept an appointment if a concern with respect to bias is raised, unless it is evident that it is solely a tactical ploy. If there is the possibility of bias, actual or perceived, it is better that he stands down even though he may feel that he could remain impartial.

If you have general concerns because you know the nominated person has had too close a relationship with the other party, then you should express your concerns both to the adjudicator and the ANB and request that he stands down to enable a different person to be nominated. If your concerns are not addressed then you should reserve your right to raise them again if justified by future events.

Concerns regarding the impartiality of the adjudicator are discussed at Part 4, section 1.3, in this book.

You may be more successful, and possibly avoid the need for legal action, if you can obtain the support of the referring party, but if persuasion does not succeed and your concerns are so serious as to justify drastic measures, then your only course of action would be to seek a court injunction restraining the adjudicator from acting. However, in the only reported case where the court was asked to replace an adjudicator (on account of his conduct following his appointment), the request was refused[26]. If during the course of the adjudication your fears materialise you could refuse to implement the decision and perhaps pre-empt the inevitable enforcement proceedings by applying to the court to have the decision set aside.

2.3   What if the referring party is in financial difficulties?

However strong you consider your defence may be to the claims being made against you, there is always the possibility that the adjudicator will find for the referring party and, in most cases, you will be faced with having to make a payment to them.

If in due course you disagree with the decision but anticipate being successful at a subsequent arbitration, will the referring party be able to repay the monies you pay out if it is in financial difficulties? So is there anything you can do at the outset of the adjudication in order to safeguard your position?

At this stage there is little you can do. The chances of persuading the adjudicator not to proceed or of obtaining an injunction to stop the adjudication are remote, particularly if the referring party can put forward some sort of argument that you are the cause of its impecuniosity because you have not been paying it properly.

The possibility of being able to halt or defer proceedings is no better if the financial position of the referring party is such that it has entered into a voluntary arrangement or even if it is actually

insolvent. The courts would be reluctant to intervene at commencement of the adjudication and you would have to wait until
after the decision and then challenge enforcement proceedings.

Challenges to the adjudicator's decision are considered later in
this Part, at section 5.

3. *Active participation*

3.1 Should you be represented?

If the dispute involves awkward contractual or legal issues it would
be wise for you to be represented. Otherwise, even though the
referring party has elected to be represented, there is no reason why
you should follow suit. If you are confident of your ability to present your case competently there is no reason why you should not
conduct your own defence to the referral, although it must be
accepted that the presence on the other side of a lawyer or professional claims consultant may be rather intimidating.

If it is only the limited availability of resources that is of concern,
you should be aware that even if you obtain outside help you will
still need to make a substantial input in explaining your case to your
representative and in researching and providing him with all the
information that he will require.

The adjudicator should maintain control of the proceedings and
not allow an aggressive representative to exert undue and improper
pressure on a party acting alone (and claims consultants are often
worse than lawyers at engaging in intimidating tactics). However, if
the dispute involves awkward legal issues then it would be safer to
seek help. Outside assistance limited to co-ordinating and
supervising basic input by your own organisation can be a good
combination.

It is worth noting that should you fall out with your adviser,
representative or expert over the fees he charges, this is not a
dispute referable to adjudication under the Act[43].

If you do elect to go it alone then it would be prudent to have a
legal or contractual expert on standby, as it were, to provide
particular advice if matters take an unexpected turn.

3.2 Agreeing an adjudicator

The other party may, sensibly, wish to agree an adjudicator with
you, in which case instead of applying to an ANB they may, at the

time they send you the notice of adjudication, suggest a particular adjudicator or they may put forward several names inviting you to make your selection. Whether you accept the name or any of the names proposed is up to you. You may be concerned that the referring party has an ulterior motive. Such concern is probably unwarranted, but you should not agree to an adjudicator who is not known to you as being experienced and competent. If you are content with the name or one of the names proposed you will need to respond promptly as the referring party will be anxious to get on with the adjudication and will not hang around waiting for you to make up your mind. However, you are entitled to ascertain the terms and conditions of the person before you agree to his acting and, if these contain anything objectionable, then give a qualified acceptance.

3.3 What if the matters in the referral notice are different from those in the notice of adjudication?

It is not uncommon for a referral notice to include issues for decision that are different from those stated in the notice of adjudication, or for new issues, which you may not have had the chance to consider, to be slipped in – perhaps accidentally but sometimes deliberately. The differences may be marginal or of no real consequence and can be overlooked. Otherwise you may wish to challenge the differences should the adjudicator himself not take issue with the referring party – he may not have spotted the differences.

Although the meaning of dispute is in itself fairly widely inter-preted, the majority of court decisions on the point have held that it is the notice of adjudication that determines what the adjudicator can deal with – see Part 2, sections 5.2 and 5.6, earlier in this book. There has been one conflicting judgment in which it was stated that the notice of adjudication and the referral notice together determine the adjudicator's jurisdiction[59]. It is suggested that the former view is to be preferred and if the referring party wishes to bring in further or different issues they can only do so with your consent.

If you do not wish the adjudicator to deal with the additional matters it is vital that you register your objection with him. If he does not agree with you, then in case you are wrong you would be wise in your response to address all matters, but it is essential that you make it clear that you are doing so without prejudice and you must reserve your right to pursue your objection at a later date. If you fail to register an objection you will be deemed to have accepted that the adjudicator has jurisdiction to decide the new or different matters.

It is not unusual for a referral notice to contain argument and evidence not previously put forward. It is quite commonplace in arbitration and litigation for material not considered prior to the action to be included in the claim or defence, and parties to adjudication have tended to do likewise. For several years the practice was accepted without protest or comment but in a case in the early part of 2002 a challenge was raised when an adjudicator took account of a report containing new argument that had not been previously advanced and considered. The court agreed that the adjudicator had not had jurisdiction to consider the report and the decision was not enforced[82].

It was considered that the judgment may have been reversed on appeal but the parties settled before the point could be considered by the higher court. If you are faced with a similar situation, as things stand you may be able to get the offending material barred by drawing attention to the judgment.

It should be noted, however, that an adjudicator can decide issues not included in the notice of adjudication if it is necessary for him to do so in order to decide issues that are included[60/62].

3.4 Your response – your defence to the referral notice

The name given to your own statement challenging the claims made in the referral notice varies, but response is most commonly used.

Unless you decide not to pursue them further, you must restate in your response any objections regarding the adjudicator or his jurisdiction that you have previously expressed. Nevertheless, you must still challenge the claims made in the referral notice but making it clear that your response is without prejudice to your opinion that (for whatever reason) the adjudication should not proceed.

The time given to you by the adjudicator in which to respond will be very limited. It should not be less than 7 days; it is unlikely to exceed 14 days. The short time available to responding parties has been the subject of considerable complaint – the ambush as it is called when a referral notice comprises several or more files of documents.

If the referral notice is voluminous, the adjudicator may direct the referring party to resubmit it in a reduced form. Alternatively, he may allow you more time and request the agreement of the referring party to an extension of the time in which he has to produce his decision. The referring party alone can agree up to 14 days. Beyond that, both parties must agree.

If you need more time than you have been allowed, tell the adjudicator you will be disadvantaged unless he gives you further time. This may cause him some difficulty on account of the limited time that he has available. He needs to allow himself sufficient time once submissions are complete so he cannot give you too long, but he will not wish to leave himself vulnerable to an accusation that he has acted unfairly by not allowing you reasonable time to respond. And, within their powers, referring parties are usually amenable to allowing the adjudicator some further time in order to ensure, as far as possible, that the dispute is given proper attention.

Your reactions should not overlook the essence of adjudication – that it is an extremely speedy if somewhat rough process which, once started, means that time is at a premium for all concerned. You must be prepared to spend long, often unsociable, hours and apply maximum resources to assembling the best response that you can in the period available to you.

You should keep your response as concise as possible, compatible with contesting everything in the referral notice with which you disagree except for trivial matters which are of no consequence. Failure to challenge will result in your being held to have accepted what the referring party has said.

If the referral notice is tidily and logically set out then it is best to follow the same format in your response. That way you will minimise the possibility of overlooking some aspect. If the submissions of the parties follow each other in this way it is very helpful to the adjudicator in tracking the arguments for and against the various issues.

If you do need to adopt a different format to that of the referral notice, perhaps because it is badly set down – and many of them are, then as far as possible your arguments should be referenced to paragraphs or headings in the referral notice. Remember you and the referring party will have lived with your contract for months or even years and will have argued the issues in dispute in correspondence and possibly across the table. The adjudicator is completely new to the situation and has to try to grasp everything in a short time. The more assistance you can give him the better will be his decision.

As with the referral notice your response must include all the documents upon which you rely. Even though a document has been included in the referral notice, it may be necessary to include it again in your response if you are relying on some different part or aspect of it than did the referring party. Do not duplicate documents unnecessarily, but inclusion in this way will assist the adjudicator to follow your train of argument. As with a good referral notice, your response documents should be readable and appropriately referenced.

Where quantum is involved it is insufficient simply to refute the claimed entitlement. You may be of the opinion that a claim item is wrong in principle but if that were not so you consider that the amount claimed is incorrect anyway. In case the adjudicator accepts the principle of the claim you should state what you consider the entitlement should then be.

If you do not state the extra cost, rate, quantity, length of delay, etc. that you consider would be appropriate should the referred claims be upheld in principle, the adjudicator may decide that he must accept the referring party's quantum because he has not been provided with an alternative. Adjudicators have the authority to work out for themselves from the evidence put before them what they consider quantum should be, but some adopt an all-or-nothing approach if they are presented with a figure from one party but nothing except a blanket denial from the other. Even if he does not accept your figures these may at least highlight to the adjudicator errors in the referring party's calculations.

You may accept that the quantum claimed is correct were the adjudicator, despite your arguments against the principle of entitlement, to agree with the referring party. It is helpful in such cases to say so.

As in respect of the referral notice, the adjudicator may ask questions of your response in order to clarify certain points or assist his understanding. He will probably state a period for you to reply and it is incumbent upon you to provide the answers or information requested on time. If the adjudicator does not stipulate a time you should reply as quickly as possible keeping in mind the period remaining in which he must give his decision.

3.5   Can you request a meeting?

Both parties have equal rights and you are entitled to request a meeting if the referring party has not done so or if the adjudicator has not himself decided a meeting would be beneficial.

Indeed, if you are faced with a lengthy submission from the

referring party and are having difficulty in compiling a comprehensive response in the degree of detail you feel necessary and in the limited time available, then it is a good tactical move to advise the adjudicator that there is argument that you wish to present orally. State that you will be prejudiced if not given the opportunity. Provided your request is made fairly promptly after receipt of the referral, the adjudicator will almost certainly agree; he will not wish to be vulnerable to an accusation of procedural unfairness.

The presentation of oral evidence to the adjudicator is discussed at Part 4, section 1.4, in this book.

3.6   Can you submit a counterclaim?

By its nature a counterclaim must be a separate dispute. The adjudicator will not therefore be permitted to take it into consideration unless it was the subject of a valid withholding notice (see Part 5 of this book) or his jurisdiction has been specifically extended to consider it.

But a counterclaim is sometimes confused with a defence. Perhaps you consider that the amount claimed by the referring party is not actually due because, for example, the work or service to which it relates has not been properly executed. Any reduction in the amount otherwise due would then be an abatement not a set-off (counterclaim). The courts have created some uncertainty by expressing opposing views on whether or not there is a difference between set-off and abatement as regards the need for a withholding notice (see Part 5 of this book).

If you have failed to give a valid withholding notice it may nevertheless be convenient to deal with your counterclaim at the same time as the referred claim because some or all of the facts are relevant to the claims of both parties, but the agreement of the other party is required before it can be considered by the adjudicator who is in place.

If you can persuade the referring party that it would be more economical and beneficial to both of you for your counterclaim to be dealt with at the same time rather than for you to instigate a separate adjudication conducted by a different adjudicator, then you can agree to extend the existing adjudicator's jurisdiction. Alternatively, you may be able to agree that your counterclaim is the subject of a separate referral conducted by the same adjudicator.

3.7   Can you settle during the course of the adjudication?

As with any legal process you have the right to reach a settlement with the other party at any time. Once you have commenced negotiations it would be prudent to agree with the referring party that the adjudicator is informed (the IChemE Adjudication Rules require this) and request him to suspend his activities pending the outcome; and if a settlement is not achieved and he then needs more time the parties will consent. There is nothing wrong with such an arrangement and it will prevent the incurrence of further abortive costs. If you reach a settlement make sure it includes a provision as to who pays the adjudicator's costs.

3.8   What if the referring party asks the adjudicator for recovery of its costs?

You may find that the notice of referral includes a request that the adjudicator orders, in the event that the referring party is successful, that you pay their costs. This is quite proper provided the operative adjudication procedure permits it, but most do not allow the adjudicator to allocate parties' costs.

The referring party may have misunderstood the position or may be trying to influence the adjudicator into accepting that he has jurisdiction to allocate costs. If you do not wish to accept the risk of costs against you, i.e. pay the referring party's costs if you lose, then you must state in your response that the request is not valid because the adjudicator does not have jurisdiction – and he will have to agree with you.

It is important to realise that if, instead of reminding the adjudicator in your response that he has no jurisdiction to award costs, you argue the point as to who should receive them, then you will likely be held to have accepted that he does indeed have jurisdiction and you will be bound by his decision.

If the procedure does not give the adjudicator the power to decide costs but you are prepared to accept the risk of paying the referring party's costs if you lose, in return for them paying yours if you win, then you should make a similar request in your response as appeared in the referral notice. In so doing you will have jointly given the adjudicator jurisdiction to make a decision on the parties' costs.

4. When the decision is ready

4.1 When do you get to see the decision?

You are entitled to receive the decision at the same time as the referring party even though the referring party may have initially paid the whole of the adjudicator's costs.

4.2 Is it necessary to contribute to the adjudicator's charges?

The decision will allocate the adjudicator's charges between the parties. How he does this will depend on the extent to which the referring party has been successful and whether one of the parties has caused him to waste time due to inadequacy of their submissions and their behaviour during the adjudication. You may be directed to pay all or part of the charges or you may not have to pay anything. Behaviour of the parties prior to the adjudication cannot be taken into account in the allocation of the adjudicator's costs.

However, if the adjudicator requires payment of his charges before he releases his decision, it is common practice, pending sight of his allocation of costs, for each party to pay one half. You may therefore be approached by the referring party requesting your agreement to share costs in the first instance. Unless you consider that the referring party has acted irresponsibly in calling for adjudication or you are concerned that you may not recover your share in the event that the adjudicator finds in your favour, then it is reasonable to agree to contribute half of the adjudicator's costs in order to secure the decision but you are under no obligation to do so.

4.3 What if the referring party does not take up the decision?

The referring party, having seen your response or perhaps as a result of some searching questions from the adjudicator, may decide to abandon its claim and is not therefore interested in seeing the decision. There may be some other reason for its reluctance.

You may not share the referring party's disinterest and, unless through the adjudication procedure or the adjudicator's terms and conditions he has a lien on his decision, you can call upon him to release it and he must comply with your request. He will then require immediate payment from either or both parties depending on how he has allocated his costs, and if payment is not made then he will sue the defaulting party or both.

The parties being jointly and severally liable for the adjudicator's reasonable costs this unfortunately means that you may end up paying the whole of them if for some reason the referring party is unable to make any payment.

Even if neither party takes up the decision the adjudicator is still entitled to his reasonable costs, but once he has received full payment he is obliged to release the decision to both parties.

4.4 What if you are unable to pay what the adjudicator has ordered?

The amount the adjudicator ordered you to pay is a debt and becomes another of your financial obligations. If it is solely your financial difficulties that are preventing you from making payment then the referring party must take the same sort of action that it would do in respect of any other monies that you owed it.

5. Challenging the decision or its enforcement

You may for some reason be of the opinion that you should not be required to comply with the adjudicator's decision. Your possible reasons for not complying were noted at Part 2, section 8, earlier in this book. Assuming you do not fall into the first two categories there, your non-compliance will be because:

(1) You disagree with the content of the decision
(2) You consider the adjudicator lacked jurisdiction
(3) You consider the adjudicator exceeded his jurisdiction
(4) You consider the adjudicator acted unfairly

Alternatively you may not dispute the decision in itself but may feel you should not make any payment because:

(5) You intend to refer the dispute to arbitration or litigation and you anticipate an award or judgment in your favour entitling you to recoup what you have had to pay as a result of the decision
(6) The referring party is in some way indebted to you
(7) You anticipate that the referring party will soon be indebted to you because you have instigated proceedings against it to

> recover a sum comparable to or in excess of the amount of the decision

and the referring party's financial state is such that you are concerned that it will not be able to pay whatever comes due to you. The consequences of future non-recovery could at least be minimised by not handing over the sum the adjudicator has ordered.

Items (1), (3) and (4) in the above list are matters of common interest and are discussed at Part 4, section 3, in this book. Items (2), (5), (6) and (7) are dealt with here.

5.1   Resisting because you consider the adjudicator lacked jurisdiction

Provided you had registered your objection at the commencement of the adjudication and made it clear that your continued participation was under protest and without prejudice to your right to raise the matter later, you are entitled to argue lack of jurisdiction of the adjudicator as your defence to enforcement proceedings. This is the most common defence to enforcement of a decision. If you are successful you will have no obligation in respect of the decision.

See Part 4, section 3.2, in this book as regards an excess of jurisdiction.

5.2   Resisting because you intend to refer the dispute to arbitration or litigation

The fact that you intend to refer the dispute to arbitration or litigation (you may even have commenced proceedings) is no basis for refusing to do whatever the decision directs. Section 108(3) of the Act is quite clear on this point – the decision is binding until finally determined by other proceedings. You therefore have an obligation to comply with it immediately.

Section 108(3) also refers to agreement as a means of finally concluding the dispute. Either or both parties may disagree with the decision but neither relish the prospect of the cost and effort of prolonging the dispute. See Part 4, section 4.1, in this book as regards settlements after the decision.

5.3   Resisting because the referring party is or will be indebted to you and is in financial difficulties

Even if you disagree with the adjudicator's decision requiring you to make a payment to the referring party, you may decide to write off the loss rather than refer the matter to arbitration in anticipation of obtaining an award in which the arbitrator, with the benefit of more time and perhaps better evidence, comes to a different conclusion from the adjudicator which would entitle you to recoup all or some of the payment you now have to make.

But if you were to refer the matter to arbitration and were successful, would you be able to recover the monies you pay following the adjudication, if the referring party is in financial difficulties? And if you have a counterclaim or have commenced proceedings to secure considered entitlements from the referring party, will it be able to pay you?

The courts have stated that, in enforcement proceedings, they will consider the financial circumstances of both parties if appropriate[54] but the outcome of such consideration is unpredictable, there having been some rather inconsistent judgments.

The fact that the referring party is experiencing financial problems – it may be trading at a loss and may not be able to repay you if you were successful at a future arbitration[103] – will not necessarily protect you from paying the amount ordered by the adjudicator[4]. However, in a similar later situation the court found differently and, whilst granting enforcement, ordered a stay conditional upon the amount of the decision being paid into court[93].

You may think that you have a stronger argument for not paying if the financial difficulties of the referring party are more severe and it has entered into a voluntary arrangement with its creditors. Think again. The existence of such an arrangement will not in itself prevent enforcement of an adjudicator's decision against you.

It is however a different matter if the referring party is insolvent, since the insolvency rules provide for mutual set-off. Adjudications are regularly raised by parties in receivership or liquidation and there have been a number of cases where losing responding parties have successfully resisted payment because the insolvency rules require any cross-claims to be considered[16/88]. A stay of an adjudicator's decision in favour of a referring party which was in administrative receivership was granted where, it was argued, the referring party would be unable to repay the amount due under the decision if subsequent proceedings in respect of a counterclaim by the responding party were successful. A condition of the stay was that the amount of the decision was paid into court[92]. A similar approach was adopted by the same judge in a different case where the financial difficulties of the referring party bordered on insolvency[122].

The courts have unfortunately exhibited an apparent inconsistency in their decisions when considering whether to give summary judgment in situations where the responding party has a considered counterclaim of some sort. To a degree this appears to stem from a variable approach by different judges to set-off and abatement, but sometimes arises from the circumstances of a particular case which are not always apparent unless the full judgment, rather than a summary report or review, is studied.

See Part 2, sections 8.2 and 8.3, earlier in this book for further discussion.

The message to paying parties who inevitably end up as responding parties is to limit their risk by giving proper attention to payment regulations, particularly the giving of a valid withholding notice, so that the merits of the set-off can be considered in the adjudication. This is not always the answer but will go a long way to minimise expense, effort and frustration.

5.4   Possible actions to resist complying with the adjudicator's decision

You could take the initiative and apply to the court for a declaration that the decision is invalid and is set aside.

Probably the better step would be not to comply with the decision and to wait for the referring party to take enforcement proceedings which would be either in the form of what is known as a statutory demand upon you or a request for summary judgment in respect of the decision. Your defence to either action would be your considered reason for non-compliance.

Part 4
Matters of Common Interest

PART 4
MATTERS OF COMMON INTEREST

Particular considerations or certain situations and problems that can arise in the course of an adjudication may be of concern to either or both parties. Sometimes a party must address the matter alone, but if both parties are affected it is best, if possible, to act together.

An approach by one party alone is likely to be dismissed as gamesmanship or an attempt to slow or frustrate the process. If both parties act in unison there is a greater chance of influencing the ANB (if it may be able to resolve whatever is giving rise to concern) or the adjudicator himself. And if your joint efforts are unsuccessful and the concern is serious enough, you can, under the Scheme paragraph 11(1), jointly revoke the adjudicator's appointment. Other procedures and rules are silent on the matter of revocation of an appointment but, adjudication being a contractual arrangement between the parties, there appears to be no reason why such power would not be implied anyway; that is, it would be deemed to have been a term of the contract.

1. *During the currency of the adjudication*

1.1 Resourcing the adjudication

What is often not appreciated until it is too late is the intensity of effort required, albeit in short bursts, during the course of an adjudication. The responding party is in the more difficult position, being faced with an often bulky referral which it has to address in a short period. Although having the advantage of being able to prepare the referral at leisure, the referring party is then faced with replying promptly to the response. Both parties may be called upon to provide further information at short notice and each has to cope with the often difficult behaviour of the other.

Even though a party may have engaged a lawyer or consultant it will still be called upon to attend consultations, research and provide information, approve submissions and so on.

It is important therefore that some arrangement is put in place to ensure that adjudication matters are given priority and that everyone who may be involved is aware of the need for a prompt reaction when called upon. Difficulties of responding promptly are aggravated if the adjudication arises near or after completion when staff who have worked on the contract have left and are no longer readily contactable. The location of documentation, if dispersed as is commonplace on large contracts, must be known and access made readily available. Of primary importance is for a system to be in place so that incoming communications are immediately given to the person co-ordinating and controlling input into the adjudication.

What all this means is that it is essential that a person be appointed to organise, co-ordinate and prepare submissions and responses. In the case of large disputes a team of several persons may be required to cope with the workload. Of equal importance is the need to set up an arrangement for prompt decision-making by someone of appropriate seniority.

Action has to be taken and decisions made often in a matter of hours rather than days, and without proper procedures in place valuable time will be lost, and input into the adjudication may be inadequate – or worse, incorrect – possibly resulting in failure rather than success.

1.2   Concerns regarding the competence of the appointed adjudicator

If you have been able to agree on an adjudicator, then clearly no problem will arise unless both parties have made a serious misjudgement. However, the standards required by ANBs of those on their lists differ and the quality of adjudicators does vary. Consequently, if you have had to request an appointment, you may find you have been given someone in whom you cannot have confidence.

The referring party is best placed to ensure that the ANB, as far as possible, appoints a suitable adjudicator. It is the referring party that provides the details of the contract and the nature of the dispute and states the sort of qualifications and experience he should have. An ANB is more likely to listen to the referring party than the responding party – even if the latter does get the opportunity of making its views known in time.

From time to time, however, someone gets appointed who should not have been. There is an understandable reluctance by a party acting alone to voice a criticism of the adjudicator for fear that his decision may reflect his displeasure. It is preferable therefore to convey your concerns to the other party as soon as possible and try to secure its support.

If there are genuine doubts as to the ability of the adjudicator or there is a risk – actual or perceived – that he might not be impartial, it would be foolish for the second party not to give serious consideration to the concerns of the first. The last thing either party should want is for the decision to be set aside and the whole process started again.

If the other party shares your concern, try to get its agreement to send a joint letter to the adjudicator stating why you consider he should not act and requesting him to stand down. Hopefully he will accept your concern but if he does not, you will have to decide whether jointly to revoke his appointment.

If the support of the other party is not forthcoming, you must decide whether to pursue your protest alone – and how far to take it.

The only reported case on the matter of the appointed adjudicator concerned a referring party faced with an adjudicator which the party had specifically advised the ANB it did not wish to be appointed. It sought an order that he be removed for that and another reason but the court refused[26].

If you are the responding party you are in an even more difficult position and, acting alone, you will probably stand little chance of getting the ANB to renominate or the adjudicator to stand down.

If despite a request from one of the parties to stand down the adjudicator continues to act, the concerned party should record that it is continuing to participate in the adjudication under protest and reserve its right to raise its concern again at a later date; this would normally be after sight of the decision.

A decision will not be set aside merely because the adjudicator has made an error, but if there is evidence that he simply did not understand what the dispute was about and ought not have agreed to act (or continued to act once he realised) he may be held to have acted in bad faith by continuing. You then stand a reasonable chance of persuading the courts not to uphold the decision. However, that is when you have actual evidence of the unsuitability of the adjudicator.

1.3   Concerns regarding the behaviour of the adjudicator

Section 108(2)(e) of the Act imposes a duty on the adjudicator to act impartially. For some time there was uncertainty as to what this meant in practical terms. In the course of the first enforcement proceedings[66] it was stated that an adjudicator's decision would be upheld even if there had been a breach of natural justice. However, this view was not upheld in subsequent cases.

The answer depended to some extent on whether adjudication was a judicial process. Eventually it was decided that it was and the rules of natural justice therefore applied[34/35]. There is no clear definition as to what these rules are but they certainly include a requirement that proceedings are conducted fairly. More recently the term procedural fairness has been coined and this is probably the best description of the manner in which an adjudication should be conducted. Even though the law itself may not be fair, the process of applying it must be.

Whatever term is applied the essential requirement of the adjudication process is that the adjudicator should provide both parties with equal opportunity to present their arguments. A number of adjudicators have incorrectly exercised the power enabling them *to take the initiative in ascertaining the facts and the law* (section 108(2)(f) of the Act) in that they have failed to ensure that both parties were aware of information they obtained[118]; and if the adjudicator has a discussion with one party he must inform the other as to what it was about[35]. A decision based on programming work done by the adjudicator himself which was not conveyed to the parties for comment was held to be invalid[9].

The position appears to be somewhat different as regards investigating the law. An adjudicator who ascertained a point of law without reference to either party was held to have acted properly[60].

Even though you may not have had any reservations about the adjudicator at his appointment, concern may subsequently arise. If in the course of the proceedings you consider that the adjudicator is not acting impartially or fairly you should advise him immediately. By so doing you will give him the opportunity there and then to address the matter – if, of course, he agrees with you. If he knows that you are unhappy as regards his behaviour he will be conscious of the possibility that you may challenge his decision on the grounds of bias, and that is not something relished by adjudicators.

The fact that you have registered your feelings will assist you if you remain concerned to the extent that you decide to challenge the decision.

If both parties share some serious concern (whatever it may be) then the adjudicator's appointment can be jointly revoked at any time prior to his making a decision. Provided he has not seriously misconducted the adjudication the parties would be responsible for meeting any reasonable costs that the adjudicator has incurred.

There is one adjudication procedure which is more helpful as regards dealing with an unsatisfactory adjudicator. Under the TeCSA Rules either party may apply to the chairman of TeCSA requesting that the adjudicator be replaced. The Rules are unique in that they give the chairman power to appoint a replacement adjudicator if he is:

- Not acting impartially
- Physically or mentally incapable of conducting the adjudication
- Failing to proceed with the adjudication with necessary despatch.

1.4   Can you be required to give evidence before the adjudicator?

It was never the intention that adjudication should include a hearing at which evidence is addressed through the parties' legal representatives, as in arbitration or litigation. However, the adjudicator is entitled to convene a meeting in order to hear representations or to receive oral evidence. Alternatively he may agree to a meeting requested by one or both parties. You cannot refuse to attend a meeting if the adjudicator directs. If you do not attend then he may draw such inferences from your absence as are justified.

The need for a meeting will be dictated by the nature and complexity of the dispute. If there is expert evidence to be compared or serious disagreement on facts, then a meeting will often be beneficial as a means of clarifying matters without the need for further written submissions.

However, some adjudicators call a meeting as a matter of course. Certainly a general discussion enables the adjudicator to obtain a better grasp of the dispute at an early stage and meeting the persons involved face to face assists in deciding between contradictory evidence. Any meeting must of course be held on a mutually convenient date although too much emphasis should not be placed on the convenience aspect. Everyone is expected to put himself out to some degree in order to accommodate the tight time-scale of the adjudication process.

Although any meeting will be less formal than is customary at arbitration, you should ensure that you and your team are properly prepared. If the adjudicator has not provided an agenda then you should request one. The adjudicator will generally direct questions himself but he may permit cross-questioning of those attending. However, he has a duty to ensure that the proceedings are conducted fairly and in an orderly manner to take account of any imbalance between the parties. There is no requirement to give evidence under oath.

You are entitled to be represented but there may be restrictions on the number of representatives – for example the Scheme, paragraph 16(2), allows only one person unless the adjudicator otherwise permits.

1.5 Should there be a site inspection?

The adjudicator may ask for a site inspection even though the dispute does not relate to workmanship. There may, for example, have been problems with access or interference by other contractors; a visit to site will give the adjudicator a better appreciation of the difficulties.

Even though the adjudicator himself does not raise the matter, either party may request that he visits the site; perhaps there are particular aspects which a party wishes to emphasise and which cannot be adequately described in writing or even by photographs. The adjudicator does not have to agree but he will normally visit if requested. He may take the opportunity to combine the site visit with a meeting at which he seeks further information from the parties or allows them to make oral submissions.

1.6 Are you required to comply with the demands of the other party?

The parties are not in a position to make demands of each other although some – or rather their representatives – behave very aggressively and try to dominate the proceedings. The adjudicator ought not to permit this sort of behaviour and he should maintain control himself.

As regards any information required from the other side, the parties ought to communicate directly with each other. It saves time and it is not necessary to channel everything through the adjudicator, although it is common practice to copy correspondence to the adjudicator so that he knows a request has been made.

You must react promptly to any reasonable request from the other party. If the adjudicator becomes aware of a request he may set a time in which you must respond. He is always conscious of the difficulties that he will encounter if time is eroded by dilatory behaviour of the parties.

Whilst you should always respond to any reasonable request, if you have received one which you consider to be improper, unnecessary or onerous you should refuse, giving your reasons, or at least ask for an explanation as to the purpose of the request. If a problem persists then either party may seek a direction from the adjudicator – to uphold or reject the request as the case may be.

If some comment is made by the other party either to you or to the adjudicator which reflects adversely upon you or which may influence him to your detriment, you should react promptly, firmly but politely if you consider it to be in some way incorrect, unjustified or unreasonable and, if necessary, you should ask the adjudicator to take any reasonable action you consider appropriate. If you are concerned at the general behaviour of the other party you should write to the adjudicator and request his intervention. He may not agree with you but you will at least have recorded your protest.

You are entitled to a reasonable time in which to respond to a request, but remember that reasonable is measured in the context of the restricted period in which the adjudicator must reach a decision.

1.7 Can you object to the directions of the adjudicator?

If the adjudicator is experienced and competent he should not give directions which are objectionable. However there will be occasions when perhaps he does not appreciate what his directions involve.

If the direction is one against which you have a fundamental objection, then you should state your reasons as to why you consider that you should not be expected to comply. If the adjudicator persists then you must make a decision either to withdraw your objection or to refuse to comply. The latter is a serious step which inevitably will influence the decision, so it may be wise to seek expert advice.

1.8 What if you are unable to comply with the directions of the adjudicator?

You should do your utmost to comply with any reasonable direction of the adjudicator, but if for some good reason you cannot, then explain immediately your difficulty to him; perhaps you do not have the information he has asked for or you need more time to provide it, but be as constructive as you can. Where appropriate offer an alternative, but whatever you do, you must not ignore the request and hope it will be forgotten.

If the adjudicator acknowledges your problem, he should react as fairly as the time constraints permit. If, however, he thinks you are not being diligent, your non-compliance will be to your detriment. Whether or not you are behaving diligently will be measured in the context of the limited duration of the adjudication process.

1.9   What if new material is introduced?

The referral notice must follow what is in the notice of adjudication and no new issues may be introduced without the consent of the responding party. (See Part 3, section 3.3, earlier in this book.) Provided that the original issues were clearly defined, an adjudicator ought to realise if new issues are introduced.

In contrast he will not realise if a submission contains material not seen before by the other party. Both parties should now be wary of putting forward new argument and evidence[82] which the other had not had the opportunity to consider prior to the referral. The position is of course the same with respect to subsequent submissions – the response, reply and any correspondence that the principal submissions generate.

If the other party introduces new material then, notwithstanding past practice, the safest course of action would be to draw it to the attention of the adjudicator and request him to direct that it be withdrawn or state that he will not take it into account. Alternatively you could register your objection but respond to the new material making it clear that you are doing so without prejudice.

You may wish to address whatever has been introduced because you think you have a better argument or evidence. In that case you should simply make an appropriate submission, but having done so you will not be able to make an objection subsequently – perhaps when you find the decision has gone against you.

Instead of allowing in the new material the adjudicator may state that he is taking no account of it. You may consider this to be unsatisfactory. You may feel that once he has seen the new material the adjudicator cannot ignore it and it is bound to influence him to some extent. This is a similar situation to that which judges and arbitrators regularly encounter and they are considered to be capable of dismissing from their minds matters which have been put before them but which, for some reason, they do not intend to take into account in reaching their awards or judgments. It may be more difficult for an adjudicator with limited judicial experience to do this, but unless there is clear indication in the decision that the supposedly rejected material has not in fact been disregarded, then there is nothing that you can do. If the adjudicator has evidently taken account of matters which he said he would not, then you should be able to persuade the court not to enforce the decision on the grounds that the adjudicator did not act impartially – but get good legal advice before you try.

1.10   If you get out of your depth can the adjudicator assist?

You may have thought you would be able to cope on your own but things have become complicated. Perhaps legal or contractual arguments which you had not anticipated have arisen, or the other party has raised procedural matters which you do not feel competent to deal with.

Undoubtedly in this situation you are in great difficulty because of the tight schedule. Seek immediate advice from someone you know to be experienced in adjudication or who is recommended to you. Those familiar with adjudication will know the urgency and hopefully will provide prompt assistance. At the same time advise the adjudicator of your predicament. You may need more time to make your response or reply, or you may want time to make a supplementary submission on whatever has arisen. You may have to pay your adviser a premium rate for him to accommodate your late cry for help, but it will be worthwhile in order to get things right. Despite all your efforts your submission may still be late, but it is probably better, within reason, to make a proper submission late rather than a wrong one which may lose you the adjudication.

Notwithstanding a contrary principle being set down in the first judgment on an adjudication decision, it is now accepted that an adjudicator has a duty to conduct the proceedings in accordance with the rules of natural justice[35]. This means he has to act as fairly to each party as the time restraints permit, ensuring that he does not favour one or the other. Hence he should do his best to accommodate a particular difficulty of one party without prejudicing the other side. He must still maintain a balance without, of course, allowing his decision to be late.

This might entail giving some limited guidance on procedural matters but the adjudicator must be extremely careful to avoid any suggestion that he is assisting either party to present its case. Perhaps by explaining in some extra detail what is required, misdirected effort could be avoided and the adjudication may progress more speedily.

1.11 Can you seek the intervention of the courts?

Most court involvement in adjudication has been by way of enforcement proceedings and to date there have been only a dozen or so reported cases when a party has sought a declaration from the court or an injunction (the comparable process in Scotland is interdict) on some procedural or jurisdictional matter. On the basis that adjudication is a temporary judgment, the courts are reluctant to interfere with the process, but when asked they have responded well to the urgency of adjudication matters and it is possible to obtain a hearing at short notice.

1.12 Can the adjudication be suspended?

If for some reason it would be sensible for progress of the adjudication to be suspended, for example if negotiations have reached a promising stage, the agreement of both parties may be required in order to secure an adequate further period for the adjudicator to give his decision should a settlement not in fact be achieved. Strictly, the agreement of the adjudicator is also required but that would be a formality in the face of consensus of the parties. The adjudicator could, if he considered it desirable, suspend proceedings of his own volition but only to the extent that his decision is not delayed beyond the operative date.

1.13 What scope does the adjudicator's expert have?

An adjudicator is entitled to engage an expert — legal or technical – to assist him. Generally he does not have to obtain the permission of the parties but he does have to advise them of his intention and he must convey to the parties any information that he obtains and which will be material to his decision. Where an adjudicator and his colleagues prepared a critical path analysis it was held that the adjudicator was, in effect, assisting one party; worse, he did not give the other party the opportunity to consider it. Consequently his decision was declared invalid[9].

An expert cannot decide – in its adjudication sense – on behalf of the adjudicator any matter unless the parties agree that he has authority to do so. Where an expert carried out work upon which the adjudicator relied, the decision was upheld because, on the facts, the court was of the opinion that the parties had agreed the extent of involvement of the expert[134].

1.14 What if the adjudicator is late in giving his decision?

The referring party may extend the 28-day period in which the adjudicator must make his decision, by up to 14 days. By agreement both parties may extend it by any period. But what happens if the adjudicator still fails to give his decision on time? There are two possible situations.

The first is where the adjudicator may have become ill, or for some other reason outside his control is unable to continue. The second is where he has been inefficient in some way and has no reasonable excuse for the delay in reaching a decision.

Most procedures cater for the situation where an adjudicator is unable to act. The parties may agree a new adjudicator or, on the application of one party, the ANB may appoint a replacement.

But where the adjudicator is at fault, adjudication rules and procedures adopt different approaches to the situation. The Scheme states that a fresh notice of adjudication may be served and the process started anew. Others may qualify this facility. The ICE Procedure, for example, states that provided the decision of the first adjudicator is notified to the parties before the dispute is referred to a new adjudicator, it stands. It is necessary to check the relevant rules or procedure to ascertain the precise position.

Of course both the parties may overlook any lateness that is not too excessive and indeed this may well be the best course of action, otherwise more time and cost will be expended in going through the process again.

If a decision against you is given late and you consider it to be a bad decision, then subject to any requirements of the adjudication procedure, you could try to use its lateness as a means of escaping from it by applying to have it set aside, but your chances of success are small.

Provided that the adjudicator is not otherwise at fault (even though in your eyes he has come to the wrong answer) and you as the applicant party would not be prejudiced in some way, the court will probably allow the decision to stand on the basis that there is no benefit to be gained in setting it aside – indeed quite the reverse in view of the time that will elapse before a new decision could be given.

1.15 Requesting reasons

The explanation of the adjudicator as to how he arrived at his decision is known as providing reasons or giving a reasoned decision.

Even though the referring party has not requested reasons, there is nothing to stop the responding party from doing so unless the operative adjudication procedure specifically prohibits this. In that event the parties will have to agree to override the prohibition and the adjudicator must then comply with their joint wishes.

See section 2.2 later in this Part for further discussion.

1.16 What if the adjudicator fails to make a decision?

Lateness in providing a decision may extend to the stage where there is a failure on the part of the adjudicator to provide a decision at all, in which case the procedure in the event of failure of the adjudicator to make a decision on time will still be appropriate.

However, the adjudicator may for some reason refuse to give a decision even though he has been validly appointed and has jurisdiction to deal with matters referred to him. He is incorrect in doing so and must use the material before him to make an appropriate decision[10], as upheld by the Appeal Court[124].

2. After the adjudicator has released his decision

2.1 What happens if you do not like what you see in the decision?

In simple terms – tough luck. No-one likes to lose, especially if they thought they had a particularly strong case, but the dispute cannot be referred again, not even for example if one party finds some new evidence that might have significantly influenced the course of the adjudication. The decision is binding until the dispute is finally determined by other proceedings. Consequently it is not possible for a second adjudicator to revise, vary or set aside the decision of the first[113].

A factor which contributes significantly to parties being discontented with decisions is that they cannot see how adjudicators have reached them. Even a reasoned decision (see the next section) does not always provide the answer but in the end it matters little how illogical you consider the decision to be, there is nothing you can do about it. The courts have enforced decisions which contain acknowledged errors or even which are based on error[15]. Unless you are able to demonstrate that the adjudicator was incompetent to deal with the dispute, or that he did not act impartially, perhaps because his behaviour during the course of the proceedings was not evenly balanced, or that the decision itself exhibits possible bias, or that the adjudicator exceeded his jurisdiction or had no jurisdiction, then the likelihood is that the decision will be upheld.

2.2 Can the adjudicator be required to explain how he reached his decision?

Basically the adjudicator's decision is a simple statement as to the extent if any to which the referring party has been successful. However, it is helpful to the parties – particularly the losing party – to know how the adjudicator has reached his decision. Most importantly, if the losing party knows why it has lost it may realise that there is no point in pursuing the dispute to arbitration or litigation in the hope of getting a more favourable result.

There are those who argue that it is not necessary for an adjudicator to provide reasons. Some go further and say it is undesirable to give reasons. Lawyers' advice to adjudicators has been to avoid doing so if possible, as giving reasons facilitates criticism of the decision and provides greater scope for raising challenges. That may be the case in respect of a reasoned arbitral award but it is only marginally true of adjudicators' decisions. Perhaps a more enlightened view is that the parties, particularly the losing party, are entitled to know the adjudicator's thinking behind his decision. However, many adjudicators follow such lawyers' advice if they are allowed to.

All adjudicators should go through a reasoning process; it requires disciplined thinking and ought to be capable of being set down in writing with little additional effort on the part of the adjudicator and at relatively little extra cost to the parties. But a reluctance to give reasons is often because the adjudicator lacks confidence in his decision. Clearly if that is the case, having to set down his reasoning would present the adjudicator with something of a problem.

A further benefit of providing reasons is that the discipline involved helps the adjudicator to avoid slips and errors, but if any are made it enables them to be spotted more readily – either by him or the parties – thus providing a better opportunity for them to be corrected.

Most procedural rules leave open the question of whether reasons should be given, placing no obligation on the adjudicator but permitting either party to require him to provide them. The exception is the CIC Rules which state that reasons shall be given unless the parties agree otherwise.

Requests for reasons are often made after the decision has been made, usually from the party which thought it would be successful and wants to know why it lost. Adjudicators have complained about this practice as it often causes inconvenience and disruption. The same is true of requests made at the last minute before the decision is due. Nevertheless there is nothing to prevent either party (where the procedure permits) from requesting reasons at any time. However, you will be more popular with the adjudicator if you make your request in the early stages of the process, either in the referral notice or the response.

Some procedural rules place restrictions on when a request for reasons may be made. For example the TeCSA rules require any request to be made within seven days of the referral; in contrast, under GC/Works/1 a request may be made up to 14 days after the decision has been notified. To prevent the inconvenience of a late request the adjudicator may, in the absence of any procedural restriction, stipulate the latest date by which one may be made; his powers to control the proceedings enable him to do this.

If reasons are requested after the decision has been made and when the adjudicator will have notified his costs, he would be entitled to require an additional appropriate payment from the requesting party.

Whether reasons are given with the decision or separately does not matter. Their significance does not change on this account.

2.3   Slips and errors in the decision

You may consider that the decision contains a slip. Although not stated in the Act or the Scheme, an adjudicator can correct clerical errors or slips. Slips are such things as multiplication errors in calculations; a clerical error may be the incorrect transposition of a calculated figure to a summary. The courts have upheld that an adjudicator may attend to such matters of his own volition or at the request of one of the parties, but have expressed the view that any action should follow quickly upon the decision[14/81].

Whilst a clerical error is normally fairly evident, difficulty can arise in distinguishing between a slip and a mistake. Slips are unintentional or accidental. Mistakes may include slips but they go further and can include deliberate actions.

It has been acknowledged that adjudication is not a refined legal process. The short time-scale alone, together with the often poor quality of evidence, makes for difficulties in the adjudicator reaching a fully considered and competent decision. Consequently the courts have accepted that adjudicators will make errors or mistakes, but have taken the view that, because adjudication does not provide the final determination of a dispute, the parties must live with this[15/131].

The position may be summarised as follows.

If an adjudicator decides the wrong question he has exceeded his jurisdiction and his decision is invalid in that respect. But if he decides the right question in the wrong way that is an error which the parties have to accept[23/131]. The former case was a reversal of a lower court judgment[22]. Not dealing with all the issues put to him is also an error which does not invalidate the decision[41].

Some adjudicators will say they can correct mistakes but they are probably blurring the dividing line between mistakes and slips. If the door is open to the correction of mistakes how far could an adjudicator go? It is easy to contemplate a situation where he would, in effect, be rethinking his decision.

Your course of action if you consider that the decision contains a slip, error or omission is to draw it to the adjudicator's attention immediately and ask him to make a correction. If there has been an indisputable slip or clerical error he will almost certainly correct it. He may first obtain comments from the other party who will, understandably, try to avoid any correction in your favour. If in the end the adjudicator does not do what you think he should, then seek expert or legal advice but be prepared to accept that you may not be able to do anything about it.

It is important to note that if you request the adjudicator to correct a considered error, then you will probably be held to have accepted that the decision is valid which could limit your scope for challenging it for some other reason[100].

2.4 What if you consider the adjudicator's costs to be excessive?

The adjudicator is entitled to reasonable payment for the work he has done.

His charging rate ought to be all-inclusive; in other words the adjudicator should not apply any uplift for overheads and so on unless his terms and conditions clearly state this – and you have accepted them. He can of course charge any travelling or hotel expenses but items like secretarial costs should have been included in his hourly rate.

Few problems appear to have arisen with respect to adjudicators' hourly rates; difficulties stem from the number of hours charged. The time expended on comparable work varies tremendously from one adjudicator to another. This may be attributable to the inefficiency of some but the parties should not have to pay for inefficiency. Whether you have to argue alone, jointly with the responding party or not at all will depend on how the costs are allocated.

When trying to resolve the matter of excessive charges you can ask the adjudicator for a make-up of the hours (although many give this anyway with their invoice), state what you consider to be appropriate, and give him the opportunity to make an adjustment. Try and reach a compromise but if negotiations are unsuccessful, then he should be paid what you consider to be a reasonable amount. It is then up to the adjudicator if he so wishes to take action to secure the shortfall.

The position is more difficult if, in order to secure his decision, the adjudicator has already been paid. If he has ordered that you pay all of his costs then you are on your own as regards any challenge. If the other party is ordered to pay part of his costs then you should endeavour to enlist its support and challenge the adjudicator together. You may be able to persuade him to abide by the valuation of an independent costs expert acceptable to both sides.

If the matter is not resolved and you feel sufficiently aggrieved, your only further course of action is to challenge the costs in court. The other party, whilst unhappy with the adjudicator's charges, may not wish to go to the lengths of a legal challenge. You cannot, of course, insist on it joining you in any legal action.

There have been two cases of a challenge to the level of an adjudicator's charges. Where the paying party considered the hourly rate to be excessive and refused to pay, the court, reversing a decision of a lower court, found the rate was reasonable[109]. And it has been held that an adjudicator is entitled to reasonable payment unless he has acted in bad faith[58].

If the other party is held totally liable for the adjudicator's costs then any excessive charging will hopefully, but not necessarily, not be your problem, but see the next section as regards joint and several liability.

2.5 Joint and several responsibility for the adjudicator's costs

Joint and several liability is a common legal term which means that each party, apart from any liability of its own, is also if necessary required to meet the other party's liability.

In adjudication it is applied to the adjudicator's costs and is a device to ensure that he gets paid. Its effect is that should one party be unable to pay the adjudicator's costs, the other party must do so and endeavour to recover from the defaulting party the amount it has paid.

The situation normally only arises when the party ordered to make payment becomes insolvent. Then the other party will have to pay up when requested by the adjudicator, and join the list of creditors.

It can also arise when the party ordered to pay refuses to do so because it considers the adjudicator's costs to be excessive. If the other party agrees then it would be beneficial to support the paying party. Together you are more likely to persuade him to reduce his costs, thus avoiding his taking recovery action which may also be directed against you.

Sometimes the party ordered to pay the adjudicator's costs just fails to do so. Then the adjudicator will need to take recovery action. He may direct this solely against the defaulting party but could sue both parties jointly and severally. Even if it is not you who should be making payment, you will have to defend the action even though the court will inevitably order the other party to pay.

The provision has been criticised by users of adjudication but, bearing in mind the periodic fragility of the construction industry, it may without it be difficult to find people prepared to act as adjudicators.

3. *Challenging the decision*

Although challenging an adjudicator's decision is, in the context of legal processes, a fairly simple step, it should not be taken unless a lawyer experienced in adjudication law agrees that you have a good case. You have at risk not only your own costs but those of the other party which you will have to meet if you are unsuccessful, the total of which may be comparable to the amount of the decision.

3.1 If you conclude that the adjudicator has not acted impartially

If having received the decision you are still concerned regarding the adjudicator's behaviour during the course of the proceedings, or if you consider that the decision itself exhibits bias that may not have been apparent earlier, you may apply to the court to have it set aside.

Partiality may be blatant and may arise through ignorance or carelessness on the part of the adjudicator, such as his obtaining from one party information that he does not pass on to the other[35]. This sort of behaviour, if established, would enable the injured party with little difficulty to get the decision set aside. However, partiality may be less obvious and may be an unconscious act of the adjudicator which is often very difficult if not impossible to prove. An appropriate test for apparent as opposed to actual bias has been said to be *whether the circumstances would lead a fair-minded observer to conclude that there was a real possibility or real danger, the two being the same, that the* [adjudicator] *was biased*[48].

You must have reasonably sound evidence to support an allegation of bias. Just because the adjudicator has in your eyes been unfair – perhaps for example in respect of the time in which he has allowed you to do something or in refusing to consider a late submission you wished to make – that does not necessarily demonstrate partiality. The Act imposes an obligation on the adjudicator to act impartially but he has a tight time-scale to adhere to and is allowed to be strict provided he is even-handed.

Attempts have been made to invoke the Human Rights Act but it has been held that it does not apply to adjudication[7/13/38].

3.2 If you consider the adjudicator has exceeded his jurisdiction

Whilst there may be no doubt as regards his authority to act, it is not uncommon for an adjudicator to go further than he should.

An excess of jurisdiction is more likely to be to the disadvantage of the responding party but the referring party may also have concerns if the adjudicator has decided matters he was not asked to in the notice of adjudication. It may be undesirable to have other matters the subject of a binding decision.

If, having drawn your concern to the attention of the adjudicator, he failed to respond to your satisfaction, then you could have sought an injunction restraining him. However, the courts prefer not to intervene and, in any case, an injunction is only temporary. Instead you will probably have notified him and the other party that you are continuing to participate in the adjudication without prejudice and, in that event, you may also have endorsed your submissions to the adjudicator to that effect.

If you are the responding party, you would refuse to attend to that part of the decision in respect of which you consider the adjudicator had no jurisdiction and would wait for the other party to take enforcement proceedings. If you are the referring party and it is important to you that the matters the adjudicator should not have dealt with are not the subject of his decision, then you would have to apply to have the offending part set aside.

If the court decides that there has been an excess of jurisdiction, it will strike out the part of the decision that is the subject of the jurisdictional excess[11/50].

3.3 If you consider the adjudicator has seriously misunderstood the dispute

Everyone can make wrong decisions and judgments – adjudicators as well as arbitrators and judges – and a wrong decision does not render it invalid (see section 2.3 earlier in this Part). The fallibility of adjudicators is something with which the parties must live. But it may be evident from the decision that the adjudicator was incompetent to deal with the dispute – he may have lacked the necessary technical understanding.

If his misunderstanding is serious enough, you could apply to the court to have the decision set aside on the grounds that the adjudicator's incompetence went well beyond the making of an error and that in continuing to act he acted in bad faith. If you are the responding party, that would be your defence in any enforcement proceedings taken against you.

As adjudication provides only a temporary solution to a dispute, a court may not be prepared to set the decision aside, but if enough is at stake it would probably be worth a try, thus avoiding having to refer the dispute to arbitration in order to obtain a more satisfactory outcome. Your argument will carry greater strength if you had already expressed your concern at the time of the appointment, as having made your point to the adjudicator, he had the opportunity to withdraw before proceedings got under way.

3.4   The process leading to a challenge of the decision or its enforcement

Any concerns you may have had will be of no consequence to you if the decision is in your favour or, despite your concerns, is something you can accept.

If for whatever reason you consider the decision is invalid in whole or in part, you should first advise the other party of your objection. The other party may acknowledge that you have a point and you may be able to get the matter resolved between you.

Alternatively there may be nothing wrong with the decision but, as the responding party, you may consider that you have good reason for not making the payment ordered by the adjudicator. In such event you should advise the referring party of your reasons; it is always possible that the referring party may accept your arguments and some sort of accommodation can be reached rather than incur further cost and effort in court proceedings.

4.   *The final solution*

The Act makes it clear at section 108(3) that a valid decision of an adjudicator is binding until the dispute is finally determined by some other process.

4.1   Settlement

The parties may of course reach a settlement which will bring matters to an ultimate conclusion. Indeed it is quite common for parties to use the decision as a basis for negotiation. Even though either or both parties may not fully agree with the decision, the conclusions of the adjudicator (particularly if he has given reasons) will often enable them to reach a settlement rather than resort to arbitration or the courts.

4.2   Arbitration or litigation

If you are the loser but are not prepared to live with the decision, you may refer the dispute to arbitration or litigation – which process will depend on what your contract states in this respect.

Sometimes there is a time limit within which you must take this step and it may be different from that for referring a dispute directly to arbitration or instigating litigation. For example, under the JCT Standard Form of Building Contract there is no time limit for referral to arbitration or legal proceedings with respect to a decision given prior to the date of issue of the final certificate, save that as with any challenge to that document, it must be within 28 days of the final certificate. If the decision is given after the final certificate, the 28 days run from the date on which the adjudicator gives his decision.

In contrast, the ICE Conditions of Contract state that the adjudicator's decision becomes final as well as binding if not referred within three months; under GC/Works/1 it becomes unchallengeable after 56 days, and under the ECC a decision is not referable after four weeks. You will not find such restrictions within the adjudication procedure itself and you will need to examine the dispute resolution clauses as a whole.

In contrast the contract may specify that a dispute must first be referred to adjudication before arbitration proceedings may commence[57]. In that situation the adjudication need not necessarily be a referral under the Act.

Referring the dispute to arbitration or litigation does not mean that the decision itself is re-examined. Instead the dispute that was the subject of the adjudication is considered from first principles as if no adjudication had taken place[29] and any procedural requirements in respect of the arbitration process must be complied with[67]. However, adjudication proceedings carry limited confidentiality and there is normally no restriction on the use of material from the adjudication in any subsequent action, but it will not normally be possible to call the adjudicator to give evidence.

If the award of the court or the tribunal is different from the decision of the adjudicator, then any necessary correction such as the repayment of monies paid as a result of the decision will obviously need to be made.

Part 5
Statutory Payment Requirements

PART 5
STATUTORY PAYMENT REQUIREMENTS

Sir Michael Latham's investigation was prompted by the difficulties being experienced principally by subcontractors but also by main contractors in receiving regular and proper payment for work done. As a result new legislation relating to payment procedures in contracts was introduced through sections 109–113 of the Housing Grants, Construction and Regeneration Act. This, together with statutory adjudication, comprised the package of reforms designed to address what were seen as major problems in the construction industry – poor cash flow and excessive time spent on dispute resolution.

The industry quickly took advantage of the opportunities presented by the Act. Initially, the majority of disputes referred related to non-compliance, or considered non-compliance, with the new payment regulations. After a while paying parties adjusted to the new regime and the incidence of disputes relating to infringement of the payment procedures has fallen, but they still comprise a significant proportion of all disputes referred.

The objective of the payment provisions is to ensure that a party entitled to payment knows when he will actually be paid and the reasons for and amount of any deduction from his considered entitlement.

There is no restriction on what a contract may state as regards intervals between payments and the period in which payments have to be made. But any contract that does not include a regime which complies with the Act is ineffective and, as with adjudication, the Scheme for Construction Contracts will apply[60/61]. The difference is that, whereas total substitution must be made in respect of non-compliant adjudication provisions, the Scheme provisions relating to payment can be adopted piecemeal to make good any defects or omissions in the contract procedure.

1. Resumé of requirements of the HGCRA (sections 109–113) and related regulations of the Scheme for Construction Contracts (Part II – Payments, paragraphs 1–12)

1.1 Act: section 109

A party is entitled to stage payments for work done unless the duration of the works is to be, or it is agreed that it will likely be, less than 45 days[21].

The parties are free to agree the amounts of the stage payments and the intervals at which they become due.

Thus the party drawing up the contract can insert whatever it wishes, however objectionable. It is up to the other party whether or not it is prepared to enter into contract on such terms; it will be too late to protest later unless it is in a position to invoke the Unfair Contract Terms Act.

Scheme: paragraph 1

Construction contracts to which the payment provisions apply are termed *relevant construction contracts* – defined at paragraph 12.

Scheme: paragraph 2

Payment provisions in most forms of contract of whatever nature include detailed rules for the computation of periodic payment entitlements.

The Scheme defines the computation of any such payments in the most general terms utilising the existing provisions of the subject contract. Claims may be included in any payment since the computation is based on the contract price which is defined at paragraph 12 as *the entire sum payable under the construction contract in respect of work.*

1.2 Act: section 110

Contracts must contain an adequate mechanism for determining what payments become due and when and must provide a final date for payment of any sums due. Again, any period between the due date and the payment date may be inserted in the contract.

The amount which the payer is intending to pay must be specified in a notice to the payee not later than five days after the date on which a payment becomes due or would have become due had the payee carried out his obligations and there was no permissible abatement or set-off in respect of any other contract between the parties.

The *basis on which the amount was calculated* must also be specified. Such wording is obviously open to wide interpretation.

It is important to note that a payer may be able to set off monies considered due under another contract. The terms of the subject contract will determine whether this is allowable.

Scheme: paragraphs 3–9

Any application for a periodic or final payment, that is a *claim by the payee*, is defined at paragraph 12 as a written notice specifying the amount of any payment considered due and the basis on which it is calculated. Thus a payer could be entitled to reject an application in whole or in part for lack of reasonable definition as to what it comprises[68].

As regards relevant construction contracts, payment becomes due on the later of:

- The expiry of seven days following the payment period
- The date the payee makes a claim.

(paragraph 4)

If the subject date is the final payment under the contract, the former period is increased from 7 to 30 days measured from completion.

(paragraph 5)

Such period, except for the smallest contracts, is clearly unrealistic bearing in mind the typical process of settling the final account – application, discussion, part rejection, part payment on account, revised application, etc – and it is evident that a sensible approach will be necessary. However payees now have some leverage, a threat of referral to adjudication, if the payer is dilatory in progressing a final account.

If the contract is not a relevant construction contract there is no provision for interim payments and the contract price becomes due on the later of:

- The expiry of 30 days following completion of the work
- The date the payee makes a claim

(paragraph 6)

Should the payee become entitled to any other payment (i.e. not covered by paragraph 2) then it becomes due on the later of:

- The expiry of seven days following the completion of the work to which the payment relates
- The date the payee makes a claim.

(paragraph 7)

The final date for payment, i.e. the date by which the payment has actually to be made, is 17 days from the date a payment becomes due.

(paragraph 8)

The requirements of the Act in respect of the notice specifying the amount the payer proposes to make – often called a section 110 notice – are reiterated.

(paragraph 9)

1.3 Act: section 111

The payer must make payment of the amount specified under section 110 by the final date for payment unless what is commonly called a withholding notice has been issued[51/113/117/130]. The notice must be in writing and repeated each time the payee makes an application for payment[107].

The notice must state the amount to be withheld, and the ground for withholding it is to be specified[85]. An intention to deduct liquidated damages is not exempt from the notice requirements[81]. If there is more than one ground, each must be specified together with the individual amount withheld on account of it. The Act does not say so but it has been held that if a payee makes an incorrect overpayment against an application one month, a withholding notice is required before it can be deducted the following month[117].

The notice must be given not later than a predetermined period, termed the prescribed period, before the final date for payment. Any period may be stated in the contract. The notice referred to at section 110 suffices as a section 111 notice provided it contains the requisite information.

If an adjudicator should decide that, although the withholding notice was in itself effective, the wrong amount was withheld, then the sum incorrectly withheld is to be paid by whichever is the later of seven days from the date of the decision and the final date for payment but for the notice. In other words, it is not to be the adjudicator's personal view as to what he considers to be a reasonable period by which any such sum he orders has to be paid.

The courts have given consideration to the meaning of *amount due*. Initially some adjudicators took the view that, in the absence of a withholding notice, they should order payment of the sum claimed, that being the *amount due*. This view was supported by two early judgments[79/113]. Since then the practice has been held to be incorrect[73/103/134]. The position was further debated somewhat inconclusively in a subsequent case[30]. Misunderstanding could have been avoided by the use of the words 'amount properly due'. It has also been held that in the absence of a withholding notice the adjudicator can investigate and determine what is the *amount due*[103].

Problems have been exacerbated by confusion arising from the relative significance of set-off and abatement. An abatement is the reduction in an amount otherwise due because, for example, some defect in the subject work has been discovered. A set-off is essentially a counterclaim of the paying party which is deducted from the amount due.

In the case of an abatement it has been held that because the full amount had not become due, no withholding notice was required[118].

Unfortunately a contrary view has been expressed and it has also been held that there is no difference as regards the withholding notice requirements between abatement and set-off[73/117]. Hence at the time of writing the position remains uncertain and advice is always to give the requisite withholding notice whether you are abating or setting off.

Scheme: paragraph 10

The prescribed period is seven days.

1.4   Act: section 112

If the net amount due, i.e. the total due less any sum properly withheld, is not paid in full by the final date for payment, the payee may suspend performance of contractual obligations provided a minimum of seven days notice has been given of intention to suspend together with reasons for so doing. Performance must resume as soon as full payment of the amount due has been made. An appropriate extension of time is made to take account of the period when work is suspended.

The contract may provide a right to suspend work for reasons other than a failure to make payment as required. In extreme situations there may be a right to determine the contract, that is, to terminate it.

Although suspension is only a temporary stoppage, it is a step that should not be undertaken lightly. There may well be problems of site security and insurance while the party suspending operations is off site, and subsequent difficulty and effort in recovering the costs of the suspension. And if the party wrongly suspending were later held to have acted improperly, it would not only be unable to recover what inevitably would be onerous costs that it had incurred, but would also be liable for the costs to the other party resulting from the delay.

Unless there is a regular and persistent failure to pay on time the better course of action would be to refer the matter to adjudication.

Scheme

The Scheme makes no reference to the right of a party to suspend performance in the event of not being paid in full by the final date for payment. Should the need arise reliance would have to be placed on the Act.

1.5 Act: section 113

One of the difficulties faced by subcontractors in the past has been the practice of contractors inserting into subcontracts what are known as pay-when-paid clauses. Contractors have argued that they should not be expected to finance their subcontractors and are perfectly entitled to withhold payment until they have received payment. Unfortunately, apart from what might be called legitimate delay in payment to the subcontractor arising out of delay in payments to the main contractor, pay-when-paid clauses have been the source of considerable abuse resulting in inordinate delays before the subcontractor receives payment.

The problem extends down the contractual chain to sub-sub-contractors and suppliers. Main contractors themselves can also suffer, although the incidence of their being in a comparable position to their subcontractors is rare.

The Act renders pay-when-paid clauses ineffective although it has no influence retrospectively[36]. However, the prohibition does not apply when the party from whom the payer is awaiting payment, normally the employer, is insolvent; there is a definition, for the purpose of the Act, of the meaning of insolvency as regards a company, a partnership or an individual[36].

If a pay-when-paid clause in a contract is rendered ineffective on account of section 113, then the parties agree other terms – they cannot be imposed unilaterally.

Contractors have adopted a variety of devices to limit the effect of the prohibition of pay- when-paid clauses, such as relating the due date for payment to a date for certification which in turn depends on certification under the main contract. Although certify-when-certified provisions are effectively pay-when-paid clauses, the practice has not to date been challenged. Indeed, the domestic form of subcontract for use with the JCT family of contracts, DOM/1, states at clause 4.23.2 that the final payment is due not later than seven days after the date of issue of the final certificate issued under the main contract conditions.

Scheme: paragraph 11

If a pay-when-paid clause is rendered ineffective and the parties cannot agree other terms for payment, then the appropriate provisions within paragraphs 3–10 of the Scheme apply.

Part 6
Comparison of the Scheme for Construction Contracts and Other Principal Adjudication Provisions

PART 6
COMPARISON OF THE SCHEME FOR CONSTRUCTION CONTRACTS AND OTHER PRINCIPAL ADJUDICATION PROVISIONS

Adjudication provisions in or incorporated into contracts set down the procedure or rules for the conduct of the adjudication. Any procedure must contain the minimum requirements of section 108(1)–(4) of the Housing Grants, Construction and Regeneration Act and, normally, these are repeated verbatim albeit at varying points in an adjudication procedure. Otherwise there is no restriction on the content of a contract procedure, but failure to include all the minimum requirements of the Act will result in the procedure being ineffective and the Scheme for Construction Contracts superseding it.

Almost all the bodies that issue standard forms of contract also produce related adjudication procedures; companies and organisations which have their own forms of contract or terms and conditions do likewise. All tend to follow the general pattern of the Scheme but they do vary and it is important to be aware of the details of the operative procedure. This is particularly so in respect of procedures drawn up by companies or organisations. Unlike the balanced documents drafted by such bodies as the JCT or ICE, they will not have been the product of debate and discussion by representatives of all the likely users, but will have been prepared with the interests of the producer at heart. Most are quite satisfactory but some contain clauses designed specifically to deter anyone who may be of a mind to instigate adjudication against the producer of the document.

Adjudication procedures for use with standard forms were first produced as loose-leaf amendments but all have now been incorporated into subsequent editions of the standard forms to which they relate.

The Act has been examined in detail in Part 2 of this book. The following is a brief resumé of the more important aspects of the Scheme – other than the section 108(1)–(4) requirements considered in Part 2 – together with summaries of the more common standard procedures sufficient to demonstrate the principal differences.

1. *The Scheme for Construction Contracts (England and Wales) Regulations 1998*

The Scheme for Construction Contracts Regulations 1998 was produced for use with any form of contract which does not contain its own procedure or in which the adjudication provisions were not compliant. The Scheme is actually the Schedule to the Regulations (Regulation 4).

Notice of adjudication

Paragraph 1   – lists the minimum content of the notice.

Paragraph 3   – the notice must accompany any request for an adjudicator to act.

Appointment of adjudicator

Paragraph 2   – sets down procedures to be adopted where a person is named in the contract but is unable or unwilling to act, and what to do if no person is named (see also paragraph 6).
  – requires a person to advise whether or not he is willing to act within two days of being requested.
  – defines an adjudicator nominating body.

Paragraph 4   – defines an adjudicator as a natural person acting in his personal capacity.

Paragraph 5   – requires an ANB to notify the referring party of its selection within five days of a request to nominate, and sets down the procedure in the event that it fails to do so.

Paragraph 9   – permits the adjudicator to resign at any time but requires him to do so if the dispute is the same or substantially the same as one previously referred and decided. If the adjudicator resigns on grounds other than those stated he may not be entitled to payment.

132

> – retains the validity of the appointment and of any decision properly made notwithstanding an objection to the appointment of a particular person.

Paragraph 11 – the adjudicator's appointment may be revoked at any time by agreement of the parties. He is entitled to be paid for work done unless the revocation is the result of default or misconduct on his part.

Referral

Paragraph 7 – following the selection of an adjudicator (by whatever means) the referring party is required within seven days of the notice of adjudication to refer the dispute to him by way of a referral notice.
– what the notice itself shall comprise is not stated but it is to be accompanied by copies of or relevant extracts from the contract and of the documents to be relied upon.
– copies of the referral submission are to be sent simultaneously to every other party to the dispute.

Jurisdiction of the adjudicator

Paragraph 8 – without the consent of the parties the adjudicator may not act at the same time in respect of more than one dispute arising out of the same contract or in respect of related disputes under different contracts.

Conduct of the adjudicator

Paragraph 12 – the adjudicator must act impartially, in accordance with the relevant terms of the contract and the applicable law and he must avoid incurring unnecessary expense.

Paragraph 13 – lists specific powers of the adjudicator to enable him to conduct the proceedings effectively and efficiently. He may:

(a) request any party to the contract to supply him with such documents as he may reasonably require including; if he so directs, any written statement from any party to the contract supporting or supplementing the referral notice and any other documents given under paragraph 7(2) [the referral notice].

(b) decide the language or languages to be used in the adjudication and whether a translation of any document is to be provided and if so by whom.

(c) meet and question any of the parties to the contract and their representatives.

(d) subject to obtaining any necessary consent from a third party or parties, make such site visits and inspections as he considers appropriate, whether accompanied by the parties or not.

(e) subject to obtaining any necessary consent from a third party or parties, carry out any tests or experiments.

(f) obtain and consider such representations and submissions as he requires and, provided he has notified the parties of his intentions, appoint experts, assessors or legal advisers.

(g) give directions as to the timetable for the adjudication, any deadlines or limits as to the length of written documents or oral representations to be complied with, and

(h) issue other directions relating to the conduct of the adjudication.

Paragraph 14 – the parties must comply with any request or direction of the adjudicator.

Paragraph 15 – if without sufficient cause a party fails to comply, the adjudicator may continue, drawing such inferences from the failure to comply as he considers are justified and reaching a decision based on the information he has. He may give weight to any late evidence.

Paragraph 16 – the parties may be assisted or represented by such advisers as they consider necessary subject only to any agreement in this respect between the parties. However, the adjudicator may limit the number of persons giving oral evidence or representations.

Paragraph 17 – the adjudicator is required to consider any relevant information given to him by the parties (subject to any direction he gives in this respect). If he obtains information himself he must make it available to the parties.

Paragraph 18 – the proceedings are confidential except to the extent that disclosure is necessary for the purpose of or in connection with the adjudication.

Paragraph 20 – the parties may by agreement extend the jurisdiction of the adjudicator or he may take into account matters under the contract which he considers are necessarily connected with the dispute.
– he may open up, revise and review any decision taken or certificate given unless the contract states that these are final and conclusive.

Final and conclusive clauses normally appear only in respect of security matters in the forms of contract for government work such as GC/Works/1, although under the JCT standard forms, for example, the final certificate can become conclusive in certain respects.

The decision

Paragraph 20 – as soon as possible after he has reached his decision the adjudicator is required to deliver a copy to the parties. If he fails to reach a decision in accordance with the time-scale, either party may serve a new notice of adjudication.
– the adjudicator may award simple or compound interest but must have regard to any terms of the contract.

Paragraph 21 – if the adjudicator does not state a time for performance of his decision then it is to be complied with immediately.

There is an error in this paragraph. Reference to delivery *in accordance with this paragraph* should read *in accordance with paragraph 19 (3).*

Paragraph 22 – if required by either party the adjudicator shall give reasons.

Costs

Paragraph 25 – the parties are jointly and severally liable for the adjudicator's reasonable fees and expenses which he may apportion between them.

1.1 The Scheme for Construction Contracts in Northern Ireland Regulations (Northern Ireland) 1999

This is the same as that for England and Wales.

1.2 The Scheme for Construction Contracts (Scotland) Regulations 1998

This is essentially the same as that for England and Wales except that at paragraph 20 it states that the adjudicator may make his decision on different aspects of the dispute at different times.

There is nothing in the England and Wales version which states that he cannot do this, and providing the adjudicator makes it clear what he is doing there seems to be no reason why a split decision should not be valid.

The Scottish version takes account of the fact that the Arbitration Act 1996 does not apply to Scotland and enforcement by that route is not therefore available. Instead a decision, as with an arbitral award, may be registered for execution in the Books of Council and Session and if either a party or the adjudicator wishes to register the decision, any other party upon being requested is required immediately to give its consent.

If the other party wishes to challenge the decision for some reason it would not give consent and the successful party would need to raise judicial review proceedings – comparable to applying for summary judgment in England and Wales.

2. *The ICE Conditions of Contract, 7th Edition and the ICE Adjudication Procedure (1997)*

The Adjudication Procedure is incorporated in the Conditions of Contract through Clause 66(6)–(8).

The ICE Conditions of Contract have always contained the concept whereby any dispute must first be referred to the engineer under the contract for his decision before arbitration can be instigated.

In order to retain this concept and still comply with the requirement of the Act that a dispute can be referred to adjudication at any time, the ICE Conditions of Contract call the initial difference or argument between the parties a *matter of dissatisfaction*.

Clause 66 (part):

(1) In order to overcome where possible the causes of disputes and in those cases where disputes are likely still to arise to facilitate their clear definition and early resolution (whether by agreement or otherwise) the following procedure shall apply for the avoidance and settlement of disputes.

(2) If at any time
   (a) the Contractor is dissatisfied with any act or instruction of the Engineer's Representative or any other person responsible to the Engineer or
   (b) the Employer or the Contractor is dissatisfied with any decision opinion direction certificate or valuation of the Engineer or with any other matter arising under or in connection with the Contract or the carrying out of the Works
   the matter of dissatisfaction shall be referred to the Engineer who shall notify his written decision to the Employer and the Contractor within one month of the reference to him.

(3) The Employer and the Contractor agree that no matter shall constitute nor be said to give rise to a dispute unless and until in respect of that matter
   (a) the time for the giving of a decision by the Engineer on a matter of dissatisfaction under Clause 66(2) has expired or the decision given is unacceptable or has not been implemented and in consequence the Employer or the Contractor has served on the other and on the Engineer a notice in writing (hereinafter called the Notice of Dispute).

This procedure has been criticised by the court as being contrary to the requirements of the Act whereby any dispute may be immediately referred to adjudication, but the judge's comments did not form part of his judgment and, although persuasive, have not therefore become law[75].

The Conditions of Contract at clause 66(6)–(8) are otherwise confined to restating the minimum requirements of the Act whilst the Procedure sets down the rules for the conduct of an adjudication.

The ICE has an Adjudicator's Agreement and Schedule which is attached to and forms part of the Procedure; appointments are made on its terms and conditions. Thus where an adjudicator is named in the contract – provided he is aware of this (which is not always the case) – or where he is appointed by the Institution, he will be deemed to have accepted the terms of the Agreement.

Where the parties have agreed the adjudicator, again, they could expect that he has accepted the terms of the Agreement provided he had been made aware that the ICE Adjudication Procedure was operative.

The parties in entering into contract will have accepted the terms of the Agreement since it is part of the Procedure which, in turn, is incorporated into the contract.

In fact there is nothing unusual or untoward in the Agreement. There is provision for the insertion of an hourly rate and an entitlement to be reimbursed disbursements, some of which most adjudicators include within their rate. The only unusual feature is the entitlement of the adjudicator to an appointment fee which is deducted from his fees and expenses when these become payable. However, it is quite common practice for the Agreement not to have been completed and signed. In that case of course no appointment fee will have been stated and it would be difficult for the adjudicator to ask for one.

The Adjudication Procedure

Notice of adjudication

Paragraph 2   – the required contents of the notice are essentially the same as under the Scheme.

Appointment of adjudicator

Paragraph 3   – where a person is named in the contract or agreed prior to the issue of the notice of adjudication, the notice is to be sent to him at the same time as to the other party together with a request for confirmation within four days that he is able and willing to act.

Where a person is not named in the contract or agreed prior to the notice of adjudication, the procedure nevertheless envisages that the parties will agree an adjudicator.

– if an adjudicator is not named or already agreed, the referring party may include with the notice the names of one or more persons who have agreed to act. The responding party may select (although the words *shall select* are used it cannot be mandatory to select from the names offered) a name within four days of the date of the notice of adjudication.

– if the above procedures do not result in an appointment, either party may within a further three days apply to any other person named in the contract or, if none is named, to the Institution for an appointment which will be an Institution not a presidential appointment.

It is interesting to note that if the responding party takes the four days allowed before it rejects a person proposed by the referring party, and the latter then takes the three days allowed before applying to the Institution for a nomination, then the whole of the seven days after the notice of adjudication for the submission of the referral notice will already have been absorbed before the Institution could even commence finding an adjudicator.

Referral

Paragraph 4 – the referral notice is called the statement of case and is to contain copies of the notice of adjudication, the contract adjudication provisions and the information relied upon including supporting documents.

It is to be sent within two days of receipt of confirmation from the adjudicator, or his selection or appointment, as the case may be, but the date of referral is that on which the adjudicator receives the referral submission.

Conduct of the adjudication

Paragraph 5 – the parties and the adjudicator may agree that he also decides matters not referred under the notice of adjudication.

– the adjudicator has complete discretion as to how to conduct the adjudication subject only to any limitations imposed by the contract (there are none in the standard form) and the Act. He is not required to observe any rule of evidence or procedure of any court.

– the responding party has 14 days in which to submit a response to the referral notice; the adjudicator may extend this period.

The Rules do not state that he may not reduce the period for a response, although such a restriction would conflict with his general power to set the timetable for the proceedings.

– other parties may be joined by consensus and there is no restriction on who they may be.

Paragraph 6 – in the event that the adjudicator fails to reach his decision on time, either party may give seven days notice of its intention to refer the dispute to a replacement adjudicator. However, if he reaches his decision before referral of the dispute to the replacement adjudicator, this is still effective provided he has notified it to the parties.

The decision

Paragraph 6 – the adjudicator may give his decision on different aspects of the dispute at different times.

It has been held that all adjudicators may do this.

– the adjudicator is not required to give reasons for his decision.

– the adjudicator is given the same powers to award interest as under the Scheme.

There is no stated requirement to give regard to any contract term relating to interest, but it is suggested that this would be implied.

– in the event that the adjudicator fails to give his decision on time, either party may give seven days' notice of its intention to refer the dispute to a replacement adjudicator. The decision is still effective if it is delivered before the dispute has been referred to the replacement adjudicator.

– the adjudicator may on his own initiative or at the request of either party correct his decision so as to remove any clerical error, mistake or ambiguity provided the initiative is taken or the request made within 14 days of notification of his decision. The adjudicator must make his corrections within seven days of a request.

Costs

Paragraph 6 – the decision may not include an award in respect of the parties' costs. If the adjudicator fails to allocate his own costs these are shared equally.
– provided the adjudicator gives notice at least seven days before it is due, he may require pre-payment of his costs before releasing his decision.

Other matters

Paragraph 7 – the Institution is given immunity in that it is not liable for any act, omission or misconduct in connection with the appointment of adjudicators.
– the parties are required to *indemnify and save harmless*, that is, safeguard the adjudicator from all claims from third parties (for example when a subcontractor has been affected by a decision relating to a contractor/employer dispute); the parties are jointly and severally liable for such protection. This is a most important inclusion from the adjudicator's standpoint as generally adjudicators do not have immunity from claims by third parties.

3. *The JCT Standard Form of Building Contract 1998, Amendment 4, 2002*

Adjudication provisions were included in Amendment 18 to the 1980 Edition published in 1988. The Amendment was incorporated into the 1998 Edition. The adjudication provisions are at clause 41A.
    There is an Adjudication Agreement. This is a simple document and contains nothing untoward.

Notice of adjudication

Clause 41A.4 – the notice is termed a notice of intention to refer a dispute to adjudication and it is just that, since it requires only that the dispute be briefly identified.

Statutory Adjudication

In the light of the emphasis given by the courts to an adjudicator's jurisdiction being defined by the notice of adjudication rather than the referral notice, this minimum requirement as to the content of the notice of intention should not be taken too literally.

Appointment of adjudicator

Clause 41A.2 – the Appendix to the Standard Form gives four potential ANBs, the RIBA, RICS, Construction Confederation (CC) and National Specialist Contractors Council (NSCC), from which one is to be selected when a contract is being prepared. An ANB is known as a nominator.

– no-one is to be nominated or agreed who will not execute the JCT Adjudication Agreement with the parties.

There is provision for an alternative clause 41A.2 where the parties wish to have an adjudicator named in the contract. In that event the adjudicator's hourly rate or lump sum fee is inserted in the Appendix to the Form of Contract.

Referral

Clause 41A.4 – referral itself (the word notice is not used) appears to be a repeat of the notice of intention, everything else being part of the accompanying documentation.

– the referral submission, if not actually delivered, is to be sent by facsimile, recorded delivery or special delivery, and if by the first method must also be posted first class. If sent by recorded delivery or special delivery it is deemed, unless proven to the contrary, to have been received 48 hours after the date of posting, subject to the exclusion of Sundays and public holidays.

– it is contemplated that an adjudicator may not be in place, and consequently the dispute not referred to him, within seven days of the notice of intention. In that case referral is to be made immediately upon appointment as evidenced by the execution of the Adjudication Agreement by the adjudicator.

Conduct of the adjudication

Clause 41A.5 – the adjudicator is required immediately to confirm the date of receipt of the referral.

The significance of this requirement is uncertain since it is deemed to have been received 48 hours after the date of posting anyway.

– the responding party is given seven days from the date of referral (presumably this means the date of receipt of the referral) in which to respond.

– if in pursuit of ascertaining the facts the adjudicator wishes to obtain information from an employee or representative of the parties he must give prior notice of his intention.

Clearly here *representative* is not meant in the same sense as the person representing a party at the adjudication.

The decision

Clause 41A.5 – the adjudicator is not obliged to give reasons for his decision.

Costs

Clause 41A.6 – each party meets its own costs except that the cost of any test or examination that the adjudicator orders may be allocated. The adjudicator's own costs are to be shared if not allocated by him.

4. The Engineering and Construction Contract, Second Edition and Option Y(UK)2 1998

The adjudication provisions are set down in an additional secondary option:Option Y(UK)2. By introducing provisions for statutory adjudication in this way, the contract can still be used on international and excluded contracts whilst, with Option Y, it is suitable for UK Contracts subject to the Act. The original non-compliant adjudication provisions are retained in the core clauses but are superseded or amended when Option Y is introduced.

The basic provisions are the simplest of all the standard forms but they introduce the concept of a matter of dissatisfaction which has been the subject of criticism as not being in accordance with the requirements of the Act in that the referral of a dispute to adjudication is delayed[88]. Supporters of the concept argue that it is up to the parties as to how they define a dispute. Here, and under the ICE Conditions, a dispute does not arise unless the matter of dissatisfaction is not resolved.

Clause 90 – if a matter of dissatisfaction arises it is to be notified within four weeks of the dissatisfied party becoming aware of it. The contractor and the project manager are required to attend a meeting within two weeks of notification to try to resolve the matter.

If the matter is not resolved within four weeks of the notice of dissatisfaction then a dispute is deemed to exist and the adjudication process may commence.

Clause 91 – if the matter of dissatisfaction is common to the main contract and subcontract, the subcontractor may attend the meeting referred to in clause 90.

Clause 92 – if the adjudicator resigns or is unable to act and the parties cannot agree a replacement within four weeks, then the person named at Part One of the Contract Data chooses a replacement.

This is the same wording as pre-1998 and it is surprising in the light of the urgency associated with statutory adjudication that there has been no reduction in the four week period before a party can request the person named in the Contract Data to choose a replacement adjudicator.

If for some reason the Scheme applies, then the adjudicator is appointed in accordance with the provisions of the Scheme (see above).

There is an Adjudicator's Contract; one existed prior to statutory adjudication but it has been modified to eliminate conflict with the requirements of the Act when used on a contract which includes Option Y(UK)2. Unfortunately it is somewhat convoluted.

There are two Forms of Agreement within the Adjudicator's Contract – one not to be used when a UK Scheme for Construction Contracts applies, the other for use when what is called the UK HGCRA Scheme for Construction Contracts 1998 applies. The latter states:

1. the Parties appoint the Adjudicator in accordance with the Scheme for Construction Contracts and with the conditions of contract and Contract Data attached to this agreement,
2. the Adjudicator accepts this appointment and undertakes to carry out his Adjudicator's duties as described in the conditions of contract and the Scheme for Construction Contracts,
3(a) References in [certain clauses of the Adjudicator's Contract] to 'the contract between the parties' are replaced by 'the contract between the parties and the Scheme for Construction Contracts'.

There is nothing in the ECC which demands the use of the Adjudicator's Contract. However, when the sitting adjudicator resigns or is unable to act, his replacement is appointed under it (core clause 92.2). Does this mean that the Scheme automatically applies even though that may not have been the intention? Notwithstanding core clause 92.2, does it only apply if the appropriate Form of Agreement is also signed? And what is the position if the Adjudicator's Contract is ignored by the adjudicator and the parties? The position is less than clear.

The Adjudicator's Contract states what he may include in his expenses; some of the items are those which most lay adjudicators would include in their hourly rate. He invoices his fee and any expenses after he has communicated his decision to the parties, and they have to pay within three weeks unless a different period is stated in the Contract Data within the Adjudicator's Contract. The adjudicator is entitled to interest if payment is late.

Notice of adjudication

Clause 90   – the party referring the dispute is known as the notifying party.
                – there are no specific requirements as to the content of the notice of intention to refer a dispute to adjudication.

Appointment of adjudicator

The ECC contemplates the appointment of an adjudicator at commencement of the contract, in which case his name is inserted by the employer at Part One of the Contract Data. There is no fall back procedure if this is not done. However, if the Adjudicator's Contract is in place, this states that the appointment is to be made in accordance with the Scheme.

Referral

Clause 90 – there are no specific requirements in respect of the submission of the notifying party – the term notice of referral is not used – except that it is to include information to be considered by the adjudicator.

Notwithstanding the lack of specification of content, the notice of intention should clearly identify the dispute and the subsequent submission must be comprehensive.

Jurisdiction of the adjudicator

Clause 91 – if a subcontract dispute is also a dispute under the contract between contractor and employer, the contractor may submit the former at the same time as the latter and the adjudicator decides the two disputes together.

Conduct of the adjudication

Clause 90 – any further information from a party to be considered by the adjudicator is to be provided within 14 days of referral.

This is rather unrealistic as the time when further information (from either party) needs to be provided will vary according to the nature and complexity of the dispute. It is suggested that no criticism could be made of the adjudicator if he ignored this provision.

The decision

Clause 90 – the adjudicator is required to give reasons.

Clause 92 – any assessment of delay or additional cost incurred by the contractor is made by the adjudicator in the same way as a compensation event is assessed.

It is important therefore that the adjudicator should be familiar with the ECC form of contract in order to understand the complexities of the compensation event procedure.

5. Institution of Chemical Engineers Form of Contract, Third Edition 2002 and Adjudication Rules, Second Edition 2001

Although originally produced for process plant construction, the IChemE form is now more widely used and is favoured by some authorities for civil engineering works.

The various versions are popularly known by their colour. The adjudication provisions within the principal forms (at clause 46 for reimbursable contracts – The Green Book) are quite brief, essentially restating section 108 of the Act. Procedure is set down in separate Adjudication Rules – The Grey Book.

The principal forms contain the concept of referring a matter of dissatisfaction to the project manager before a dispute arises, but that step may be omitted on contracts subject to the Act.

The Adjudication Rules

Notice of adjudication

Rule 3   – nothing is specified in the Rules as to what the notice should contain but a standard form of notice is at Annex A to the Rules.

Appointment

Rule 3   – an adjudicator may be named in the contract. If not, the referring party may include names with the notice of adjudication. Failing agreement between the parties within four days, an application for an appointment may be made to the president of the IChemE.

Rule 4   – upon selection the adjudicator informs the parties in writing of his willingness to act. He and the parties sign the Agreement at Annex B. The return of the signed Agreement to the adjudicator within seven days confirms his appointment.

Rule 6   – the procedure contemplates only the appointment of the adjudicator within seven days of the notice of adjudication. Referral of the dispute to him is to be in accordance with his directions.

Referral

Rule 6   – unusually, instead of stating directly what the referral – called a statement – is to include, this Rule defines what the adjudicator shall direct that it is to include. As regards any witnesses (to fact or opinion), only the information as to the subject to be covered, rather than a full statement or report, is required.
      – the date of referral is the date on which the adjudicator receives the referring party's submission.

Conduct of the adjudication

Rule 6   – the adjudicator is to give a similar direction to the responding party as to the referring party.
      – continued negotiation is encouraged but if the parties engage in discussion then any conclusions reached are to be notified to the adjudicator in an agreed written account. If partial agreement is reached, the adjudicator is to be advised of any issues no longer in dispute.

The decision

Rule 7   – in addition to the usual powers, the adjudicator is specifically empowered to give a decision on different aspects of the dispute at different times.

Rule 8   – he is to give reasons unless the parties agree otherwise.
      – even though the decision is late, it is still valid provided it is delivered before the appointment of a replacement adjudicator.

Costs

Rule 8   – each party bears its own costs and pays half of the adjudicator's costs unless otherwise ordered.

> – the adjudicator may notify the parties not less than seven days before his decision is due that he requires payment or part payment of his costs before release.

Annex D – suggests how the adjudicator's costs may be compiled.
   – there is provision for the adjudicator to insert a daily rate of interest chargeable should he not receive payment within 15 days of the date of his invoice.
   – where the adjudication becomes prolonged the adjudicator may invoice monthly.

6.   GC/Works/1 (1998) – General Conditions

The predecessors to this form of contract were designed specifically for government works and were produced by the Property Services Agency. That task was taken over by the Property Advisers to the Civil Estate (PACE).

Adjudication provisions are contained within the conditions at clause 59.

Notice of Adjudication

Clause 59(1)   – the contents of the notice are not specified.

Appointment of adjudicator

Clause 59(3)   – it is the intention that an adjudicator be named in the contract, and the Abstract of Particulars provides also for naming a substitute adjudicator should the first name for some reason be unable to act.
   – there is a prescribed form of appointment (referred to in the Abstract of Particulars) and this should be completed within 28 days of the acceptance of tender. If appointment of one of the named adjudicators is not achieved and agreement cannot be reached, then either party may apply to the Chartered Institute of Arbitrators (or the Scottish branch of the Institute for Scottish Contracts) for an appointment. The prescribed

149

form of appointment is to be used so far as is reasonably practicable.

Referral

Clause 59(2)   – the notice of referral is to set out the principal facts and arguments relating to the dispute and is to enclose copies of all relevant documents in the possession of the referring party. Copies of the submission are also to be sent to the project manager and quantity surveyor.

Conduct of the adjudication

Clause 59(4)   – the project manager or the quantity surveyor may make representations to the adjudicator.
There appears to be nothing to prevent either person actually representing the employer but the provision introduces possible third party involvement and as such is unique among the common adjudication procedures.

Clause 59(5)   – the procedure is also unique in preventing the adjudicator from delivering his decision in less than ten days from receipt of the referral.

Clause 59(6)   – the adjudicator is stated to have all the powers of an arbitrator.

The decision

Clause 59(5)   – the decision is still valid even if issued late.
This provision is contrary to section 108(2)(c) of the Act. The validity of a late decision has not as yet been tested in the courts but it is questionable as to the extent this provision would be upheld.

Clause 59(6)   – the adjudicator has unrestricted power to award interest, not controlled by any limitations in the contract clause dealing with finance charges.

Clause 59(8)   – in respect of disputes arising out of certain decisions of the employer, regarding:
  • site administration
  • passes

- assignment
- determination in various circumstances

the adjudicator is not empowered to vary or override the decision, the only remedy being financial compensation.

These restrictions reflect the finality and conclusivity of certain decisions which have always been a feature of GC/Works/1 and its predecessor CCC/Works/1, necessitated primarily by security considerations in respect of work executed on government establishments.

Clause 59(10) – either party may request reasons.

Costs

Clause 59(6) – the adjudicator may assess and award legal and other costs.

7. *The Technology and Construction Solicitors Association 2002 Version 2.0 Procedural Rules for Adjudication*

These Rules are substantially the same as those produced initially by TeCSA's predecessor, the Official Referees Solicitors Association.

They have been subject to several amendments and are distinguished from other rules and procedures in that they contain a number of unique provisions.

Notice of adjudication

Rule 3 – No details as to content are specified other than that the dispute is to be identified in general terms.

It is recommended that this rather loose provision should not be interpreted literally.

Appointment of adjudicator

Rule 5 – failing agreement an appointment is made by the chairman of TeCSA. There is no provision for nomination by an ANB.

– the difficulties that may arise in securing an adjudicator are recognised in that the chairman will *endeavour* to do so such that the dispute may be referred to the adjudicator within seven days of the notice requiring adjudication.

– if there is any doubt as to the readiness or willingness of a named adjudicator to act, then a chairman-appointed adjudicator acts.

Rule 6   – the adjudicator (whether agreed or appointed) is required to confirm his acceptance within seven days of the notice.

Clearly, unless the adjudicator confirms much earlier than seven days, the chairman will be unable to appoint in time to enable referral within seven days of the notice.

Rule 8   – the chairman may replace the appointed adjudicator by another nominated person if and when it appears necessary for him to do so.

Necessity may arise if any party represents that the adjudicator is not acting impartially or is physically or mentally incapable of conducting the adjudication or is not proceeding with necessary despatch or making his decision.

Referral

Rule 7   – the date when the adjudicator receives the referral notice, which he is required to confirm to the parties, is the date of referral.

The contents of the referral notice are not specified.

Jurisdiction of the adjudicator

Rule 9   – there is no bar to the adjudicator being appointed to more than one dispute arising out of the same contract.

Rule 13   – the scope of the adjudication may be extended by agreement of the parties to include matters beyond those identified in the notice of adjudication. The adjudicator has the discretion to decide other matters in order that the adjudication may be effective and/or meaningful.

Rule 14   – in contrast to all other rules and procedures the adjudicator may decide on his own substantive jurisdiction.

If he does so his decision on jurisdiction is treated in the same way as any other aspect of the decision, making it extremely difficult to challenge.

Conduct of the adjudication

Rule 21   – in addition to the usual wide range of powers given to him, the adjudicator may facilitate an agreement between the parties.

There is nothing to say that an adjudicator under any other procedure cannot do this but, once he has tried, his position may well have been prejudiced if he has not been successful[48]. The position would be no different under these Rules.

Rule 23   – either party has a right to a hearing or meeting
     – the adjudicator cannot require an advance payment or security for his fees.

This would include requiring payment before he releases his decision.

The decision

Rule 17   – the adjudicator is required to endeavour to ensure that his decision reflects the legal entitlements of the parties. If time restraints prevent him reaching a concluded view, then his decision should reflect his fair and reasonable view in the light of the facts and the law as he has been able to ascertain them.

This provision highlights the problems faced by adjudicators in resolving complicated factual or legal arguments within the time-scale, and the acceptance by the courts that, consequently, mistakes will be made.

Rule 30   – the adjudicator may award simple or compound interest.

Rule 31   – the adjudicator is required to give reasons if requested by any party within seven days of referral.

Costs

Rule 27   – the adjudicator's own fees shall not exceed £1250 per day plus expenses.

Rule 28   – if the parties agree, the adjudicator may allocate their costs.

Rule 29   – notwithstanding anything to the contrary in the contract between the parties, the adjudicator cannot order the referring party to pay the costs of any other party solely by reason of having referred the dispute to adjudication.

         Thus this rule overcomes an extremely objectionable provision which appears in some non-standard adjudication provisions.

Other matters

Rule 32   – there is the normal provision enabling slips and omissions to be corrected, but whereas the adjudicator acting on his own initiative must make a correction as soon as possible after he becomes aware of the need to do so, any request from a party must be made within five days of the date the decision is delivered – unless the adjudicator has specified a shorter period.

Rule 34   – the immunity of the adjudicator and his employees or agent is extended to include TeCSA and its chairman and deputy, unless there has been an act of bad faith.

In contrast to other rules and procedures the TeCSA Rules endeavour to control any involvement of the courts:

Rule 33   – the parties are entitled to summary enforcement of the decision regardless of whether it is the subject of challenge or review.

         The adjudication process is completed (save for correction of slips) once a decision has been made. It is suggested that enforcement is a matter for the courts to decide. If there is a challenge, inevitably this is heard before summary judgment is given.

         An amendment by a party to an earlier version

of the Rules (The ORSA Rules), to the effect that no application to a court in respect of the conduct of the adjudication or the adjudicator's decision could be made until after completion of the contract, was held to be unsustainable[39].

– no party is entitled to raise any right of set-off, counterclaim or abatement in connection with enforcement proceedings.

After several years of some uncertainty as to the right to set off against an adjudicator's decision, particularly where there is an apparently over-riding contract clause, a Court of Appeal decision has upheld the principle of this Rule[132].

The Rule may however be ineffective where there is insolvency. (See Part 2, sections 8.2 and 8.3, and Part 3, section 5.3 earlier in this book.)

Rule 38 – the conduct of the adjudicator or his decision cannot be challenged until he has made or refused to make his decision and until the applicant party has complied with any decision.

It is possible that the courts, if asked, would, depending on the circumstances, take a different view albeit they are reluctant to interfere with the process of adjudication and generally prefer not to become involved until after a decision has been made.

8. Construction Industry Council Model Adjudication Procedure, Second Edition

Produced by the Construction Industry Council, the Model Adjudication Procedure is used primarily in service contracts between employers and their designers and advisers.

There is an Agreement which contains nothing unusual or untoward. It provides for the insertion of an hourly rate and a maximum daily rate.

Notice of adjudication

Rule 8 – the notice is to include a brief statement of the issue or issues to be referred.

This wording recognised, prior to the judgment that verified the point, that a dispute may comprise more than one issue.

Referral

Rule 14   – recognising the importance of the wording of the notice of adjudication, the Rules state that the referral – known as the statement of case – is to be confined to the issues raised in the notice of adjudication.

Rule 15   – the date of referral is that on which the adjudicator receives the statement of case, which is to contain the usual information.

Appointment of adjudicator

Rule 10   – if the contract does not name an adjudicator and the parties cannot agree, the nominating body is the Construction Industry Council.

Rule 12   – adoption of the rules means the appointment is in the terms and conditions set down in an attached Agreement unless the subject contract states otherwise.

Jurisdiction of the adjudicator

Rule 22   – other parties may be joined in the adjudication by agreement of the parties and the adjudicator.

Conduct of the adjudication

Rule 17   – the adjudicator is not required to observe any rule of evidence, procedure or otherwise of any court or tribunal; this would be implied into other adjudication procedures if not stated.

Rule 20   – other matters may be included within the scope of the adjudication subject to the agreement of the parties and the adjudicator.

The decision

Rule 24   – the adjudicator is required to give reasons unless both parties agree that he shall not.

Rule 27   – the adjudicator may award simple or compound interest as he considers appropriate.

Costs

Rule 24   – the adjudicator is entitled to payment of his costs before releasing his decision.

Rule 28   – the parties bear their own costs.

Other matters

Rule 34   – the decision is not to be relied upon by third parties and the adjudicator has no duty of care to them.

                 As third parties are not party to the subject contract and not, therefore, to the Rules, it is questionable whether such protection would be effective.

9. Subcontract forms

Most standard forms of contract have related subcontract conditions. The adjudication provisions in a main contract form are carried down essentially unaltered to the corresponding subcontract conditions.

Appendix 1
The Housing Grants, Construction and Regeneration Act 1996

APPENDIX 1
HOUSING GRANTS, CONSTRUCTION AND REGENERATION ACT 1996

PART II

CONSTRUCTION CONTRACTS

Introductory provisions

104.–(1) In this Part a "construction contract" means an agreement with a person for any of the following–

(a) the carrying out of construction operations;

(b) arranging for the carrying out of construction operations by others, whether under sub-contract to him or otherwise;

(c) providing his own labour, or the labour of others, for the carrying out of construction operations.

Construction contracts.

(2) References in this Part to a construction contract include an agreement–

(a) to do architectural, design, or surveying work, or

(b) to provide advice on building, engineering, interior or exterior decoration or on the laying-out of landscape,

in relation to construction operations.

(3) References in this Part to a construction contract do not include a contract of employment (within the meaning of the Employment Rights Act 1996).

(4) The Secretary of State may by order add to, amend or repeal any of the provisions of subsection (1), (2) or (3) as to the agreements which are construction contracts for the purposes of this Part or are to be taken or not to be taken as included in references to such contracts.

No such order shall be made unless a draft of it has been laid before and approved by a resolution of each House of Parliament.

(5) Where an agreement relates to construction operations and other matters, this Part applies to it only so far as it relates to construction operations.

An agreement relates to construction operations so far as it makes provision of any kind within subsection (1) or (2).

(6) This Part applies only to construction contracts which–
- (a) are entered into after the commencement of this Part, and
- (b) relate to the carrying out of construction operations in England, Wales or Scotland.

(7) This Part applies whether or not the law of England and Wales or Scotland is otherwise the applicable law in relation to the contract.

Meaning of "construction operations".

105.–(1) In this Part "construction operations" means, subject as follows, operations of any of the following descriptions–
- (a) construction, alteration, repair, maintenance, extension, demolition or dismantling of buildings, or structures forming, or to form, part of the land (whether permanent or not);
- (b) construction, alteration, repair, maintenance, extension, demolition or dismantling of any works forming, or to form, part of the land, including (without prejudice to the foregoing) walls, roadworks, power-lines, telecommunication apparatus, aircraft runways, docks and harbours, railways, inland waterways, pipe-lines, reservoirs, water-mains, wells, sewers, industrial plant and installations for purposes of land drainage, coast protection or defence;
- (c) installation in any building or structure of fittings forming part of the land, including (without prejudice to the foregoing) systems of heating, lighting, air-conditioning, ventilation, power supply, drainage, sanitation, water supply or fire protection, or security or communications systems;
- (d) external or internal cleaning of buildings and structures, so far as carried out in the course of their construction, alteration, repair, extension or restoration;
- (e) operations which form an integral part of, or are preparatory to, or are for rendering complete, such operations as are previously described in this subsection, including site clearance, earth-moving, excavation, tunnelling and boring, laying of foundations, erection, maintenance or dismantling of scaffolding, site restoration, landscaping and the provision of roadways and other access works;
- (f) painting or decorating the internal or external surfaces of any building or structure.

(2) The following operations are not construction operations within the meaning of this Part–
- (a) drilling for, or extraction of, oil or natural gas;
- (b) extraction (whether by underground or surface working) of minerals; tunnelling or boring, or construction of underground works, for this purpose;
- (c) assembly, installation or demolition of plant or machinery, or erection or demolition of steelwork for the purposes of sup-

porting or providing access to plant or machinery, on a site where the primary activity is–

(i) nuclear processing, power generation, or water or effluent treatment, or

(ii) the production, transmission, processing or bulk storage (other than warehousing) of chemicals, pharmaceuticals, oil, gas, steel or food and drink;

(d) manufacture or delivery to site of–

(i) building or engineering components or equipment,

(ii) materials, plant or machinery, or

(iii) components for systems of heating, lighting, air-conditioning, ventilation, power supply, drainage, sanitation, water supply or fire protection, or for security or communications systems,

except under a contract which also provides for their installation;

(e) the making, installation and repair of artistic works, being sculptures, murals and other works which are wholly artistic in nature.

(3) The Secretary of State may by order add to, amend or repeal any of the provisions of subsection (1) or (2) as to the operations and work to be treated as construction operations for the purposes of this Part.

(4) No such order shall be made unless a draft of it has been laid before and approved by a resolution of each House of Parliament.

106.–(1) This Part does not apply–   *Provisions not applicable to contract with residential occupier.*

(a) to a construction contract with a residential occupier (see below), or

(b) to any other description of construction contract excluded from the operation of this Part by order of the Secretary of State.

(2) A construction contract with a residential occupier means a construction contract which principally relates to operations on a dwelling which one of the parties to the contract occupies, or intends to occupy, as his residence.

In this subsection "dwelling" means a dwelling-house or a flat; and for this purpose–

"dwelling-house" does not include a building containing a flat; and

"flat" means separate and self-contained premises constructed or adapted for use for residential purposes and forming part of a building from some other part of which the premises are divided horizontally.

(3) The Secretary of State may by order amend subsection (2).

(4) No order under this section shall be made unless a draft of it has been laid before and approved by a resolution of each House of Parliament.

163

Provisions applicable only to agreements in writing.

107.–(1) The provisions of this Part apply only where the construction contract is in writing, and any other agreement between the parties as to any matter is effective for the purposes of this Part only if in writing.

The expressions "agreement", "agree" and "agreed" shall be construed accordingly.

(2) There is an agreement in writing–
- (a) if the agreement is made in writing (whether or not it is signed by the parties),
- (b) if the agreement is made by exchange of communications in writing, or
- (c) if the agreement is evidenced in writing.

(3) Where parties agree otherwise than in writing by reference to terms which are in writing, they make an agreement in writing.

(4) An agreement is evidenced in writing if an agreement made otherwise than in writing is recorded by one of the parties, or by a third party, with the authority of the parties to the agreement.

(5) An exchange of written submissions in adjudication proceedings, or in arbitral or legal proceedings in which the existence of an agreement otherwise than in writing is alleged by one party against another party and not denied by the other party in his response constitutes as between those parties an agreement in writing to the effect alleged.

(6) References in this Part to anything being written or in writing include its being recorded by any means.

Adjudication

Right to refer disputes to adjudication.

108.–(1) A party to a construction contract has the right to refer a dispute arising under the contract for adjudication under a procedure complying with this section.

For this purpose "dispute" includes any difference.

(2) The contract shall–
- (a) enable a party to give notice at any time of his intention to refer a dispute to adjudication;
- (b) provide a timetable with the object of securing the appointment of the adjudicator and referral of the dispute to him within 7 days of such notice;
- (c) require the adjudicator to reach a decision within 28 days of referral or such longer period as is agreed by the parties after the dispute has been referred;
- (d) allow the adjudicator to extend the period of 28 days by up to 14 days, with the consent of the party by whom the dispute was referred;
- (e) impose a duty on the adjudicator to act impartially; and

164

(f) enable the adjudicator to take the initiative in ascertaining the facts and the law.

(3) The contract shall provide that the decision of the adjudicator is binding until the dispute is finally determined by legal proceedings, by arbitration (if the contract provides for arbitration or the parties otherwise agree to arbitration) or by agreement.

The parties may agree to accept the decision of the adjudicator as finally determining the dispute.

(4) The contract shall also provide that the adjudicator is not liable for anything done or omitted in the discharge or purported discharge of his functions as adjudicator unless the act or omission is in bad faith, and that any employee or agent of the adjudicator is similarly protected from liability.

(5) If the contract does not comply with the requirements of subsections (1) to (4), the adjudication provisions of the Scheme for Construction Contracts apply.

(6) For England and Wales, the Scheme may apply the provisions of the Arbitration Act 1996 with such adaptations and modifications as appear to the Minister making the scheme to be appropriate.

For Scotland, the Scheme may include provision conferring powers on courts in relation to adjudication and provision relating to the enforcement of the adjudicator's decision.

Payment

109.–(1) A party to a construction contract is entitled to payment by instalments, stage payments or other periodic payments for any work under the contract unless–

> Entitlement to stage payments.

    (a) it is specified in the contract that the duration of the work is to be less than 45 days, or

    (b) it is agreed between the parties that the duration of the work is estimated to be less than 45 days.

(2) The parties are free to agree the amounts of the payments and the intervals at which, or circumstances in which, they become due.

(3) In the absence of such agreement, the relevant provisions of the Scheme for Construction Contracts apply.

(4) References in the following sections to a payment under the contract include a payment by virtue of this section.

110.–(1) Every construction contract shall–

> Dates for payment.

    (a) provide an adequate mechanism for determining what payments become due under the contract, and when, and

    (b) provide for a final date for payment in relation to any sum which becomes due.

The parties are free to agree how long the period is to be between the date on which a sum becomes due and the final date for payment.

(2) Every construction contract shall provide for the giving of notice by a party not later than five days after the date on which a payment becomes due from him under the contract, or would have become due if–
- (a) the other party had carried out his obligations under the contract, and
- (b) no set-off or abatement was permitted by reference to any sum claimed to be due under one or more other contracts,

specifying the amount (if any) of the payment made or proposed to be made, and the basis on which that amount was calculated.

(3) If or to the extent that a contract does not contain such provision as is mentioned in subsection (1) or (2), the relevant provisions of the Scheme for Construction Contracts apply.

Notice of intention to withhold payment.

111.–(1) A party to a construction contract may not withhold payment after the final date for payment of a sum due under the contract unless he has given an effective notice of intention to withhold payment.

The notice mentioned in section 110(2) may suffice as a notice of intention to withhold payment if it complies with the requirements of this section.

(2) To be effective such a notice must specify–
- (a) the amount proposed to be withheld and the ground for withholding payment, or
- (b) if there is more than one ground, each ground and the amount attributable to it,

and must be given not later than the prescribed period before the final date for payment.

(3) The parties are free to agree what that prescribed period is to be.

In the absence of such agreement, the period shall be that provided by the Scheme for Construction Contracts.

(4) Where an effective notice of intention to withhold payment is given, but on the matter being referred to adjudication it is decided that the whole or part of the amount should be paid, the decision shall be construed as requiring payment not later than–
- (a) seven days from the date of the decision, or
- (b) the date which apart from the notice would have been the final date for payment,

whichever is the later.

Right to suspend performance for non-payment.

112.–(1) Where a sum due under a construction contract is not paid in full by the final date for payment and no effective notice to withhold payment has been given, the person to whom the sum is due has the right (without prejudice to any other right or remedy) to suspend performance of his obligations under the contract to the party by whom payment ought to have been made ("the party in default").

(2) The right may not be exercised without first giving to the party in default at least seven days' notice of intention to suspend performance, stating the ground or grounds on which it is intended to suspend performance.

(3) The right to suspend performance ceases when the party in default makes payment in full of the amount due.

(4) Any period during which performance is suspended in pursuance of the right conferred by this section shall be disregarded in computing for the purposes of any contractual time limit the time taken, by the party exercising the right or by a third party, to complete any work directly or indirectly affected by the exercise of the right.

Where the contractual time limit is set by reference to a date rather than a period, the date shall be adjusted accordingly.

113.–(1) A provision making payment under a construction contract conditional on the payer receiving payment from a third person is ineffective, unless that third person, or any other person payment by whom is under the contract (directly or indirectly) a condition of payment by that third person, is insolvent.

Prohibition of conditional payment provisions.

(2) For the purposes of this section a company becomes insolvent–
   (a) on the making of an administration order against it under Part II of the Insolvency Act 1986,
   (b) on the appointment of an administrative receiver or a receiver or manager of its property under Chapter I of Part III of that Act, or the appointment of a receiver under Chapter II of that Part,
   (c) on the passing of a resolution for voluntary winding-up without a declaration of solvency under section 89 of that Act, or
   (d) on the making of a winding-up order under Part IV or V of that Act.

(3) For the purposes of the section a partnership becomes insolvent–
   (a) on the making of a winding-up order against it under any provision of the Insolvency Act 1986 as applied by an order under section 420 of that Act, or
   (b) when sequestration is awarded on the estate of the partnership under section 12 of the Bankruptcy (Scotland) Act 1985 or the partnership grants a trust deed for its creditors.

(4) For the purposes of this section an individual becomes insolvent–
   (a) on the making of a bankruptcy order against him under Part IX of the Insolvency Act 1986, or

(b) on the sequestration of his estate under the Bankruptcy (Scotland) Act 1985 or when he grants a trust deed for his creditors.

(5) A company, partnership or individual shall also be treated as insolvent on the occurrence of any event corresponding to those specified in subsection (2), (3) or (4) under the law of Northern Ireland or of a country outside the United Kingdom.

(6) Where a provision is rendered ineffective by subsection (1), the parties are free to agree other terms for payment.

In the absence of such agreement, the relevant provisions of the Scheme for Construction Contracts apply.

Supplementary provisions

The Scheme for Construction Contracts.

114.–(1) The Minister shall by regulations make a scheme ("the Scheme for Construction Contracts") containing provision about the matters referred to in the preceding provisions of this Part.

(2) Before making any regulations under this section the Minister shall consult such persons as he thinks fit.

(3) In this section "the Minister" means–
(a) for England and Wales, the Secretary of State, and
(b) for Scotland, the Lord Advocate.

(4) Where any provisions of the Scheme for Construction Contracts apply by virtue of this Part in default of contractual provision agreed by the parties, they have effect as implied terms of the contract concerned.

(5) Regulations under this section shall not be made unless a draft of them has been approved by resolution of each House of Parliament.

Service of notices, &c.

115.–(1) The parties are free to agree on the manner of service of any notice or other document required or authorised to be served in pursuance of the construction contract or for any of the purposes of this Part.

(2) If or to the extent that there is no such agreement the following provisions apply.

(3) A notice or other document may be served on a person by any effective means.

(4) If a notice or other document is addressed, pre-paid and delivered by post–
(a) to the addressee's last known principal residence or, if he is or has been carrying on a trade, profession or business, his last known principal business address, or
(b) where the addressee is a body corporate, to the body's registered or principal office,
it shall be treated as effectively served.

(5) This section does not apply to the service of documents for the purposes of legal proceedings, for which provision is made by rules of court.

(6) References in this Part to a notice or other document include any form of communication in writing and references to service shall be construed accordingly.

116.–(1) For the purposes of this Part periods of time shall be reckoned as follows.

Reckoning periods of time.

(2) Where an act is required to be done within a specified period after or from a specified date, the period begins immediately after that date.

(3) Where the period would include Christmas Day, Good Friday or a day which under the Banking and Financial Dealings Act 1971 is a bank holiday in England and Wales or, as the case may be, in Scotland, that day shall be excluded.

117.–(1) This Part applies to a construction contract entered into by or on behalf of the Crown otherwise than by or on behalf of Her Majesty in her private capacity.

Crown application.

(2) This Part applies to a construction contract entered into on behalf of the Duchy of Cornwall notwithstanding any Crown interest.

(3) Where a construction contract is entered into by or on behalf of Her Majesty in right of the Duchy of Lancaster, Her Majesty shall be represented, for the purposes of any adjudication or other proceedings arising out of the contract by virtue of this Part, by the Chancellor of the Duchy or such person as he may appoint.

(4) Where a construction contract is entered into on behalf of the Duchy of Cornwall, the Duke of Cornwall or the possessor for the time being of the Duchy shall be represented, for the purposes of any adjudication or other proceedings arising out of the contract by virtue of this Part, by such person as he may appoint.

Appendix 2
The Construction Contracts (England and Wales) Exclusion Order 1998

APPENDIX 2
THE CONSTRUCTION CONTRACTS (ENGLAND AND WALES) EXCLUSION ORDER 1998

Statutory Instrument 1998 No. 648

The Secretary of State, in exercise of the powers conferred on him by sections 106(1)(b) and 146(1) of the Housing Grants, Construction and Regeneration Act 1996 and of all other powers enabling him in that behalf, hereby makes the following Order, a draft of which has been laid before and approved by resolution of, each House of Parliament:

Citation, commencement and extent

1.–(1) This Order may be cited as the Construction Contracts Exclusion Order 1998 and shall come into force at the end of the period of 3 weeks beginning with the day on which it is made ("the commencement date").

(2) This Order shall extend to England and Wales only.

Interpretation

2. In this Order, "Part II" means Part II of the Housing Grants, Construction and Regeneration Act 1996.

Agreements under statute

3. A construction contract is excluded from the operation of Part II if it is–

    (a) an agreement under section 38 (power of highway authorities to adopt by agreement) or section 278 (agreements as to execution of works) of the Highways Act 1980;

    (b) an agreement under section 106 (planning obligations), 106A (modification or discharge of planning obligations) or 299A (Crown planning obligations) of the Town and Country Planning Act 1990;

    (c) an agreement under section 104 of the Water Industry Act 1991 (agreements to adopt sewer, drain or sewage disposal works); or

(d) an externally financed, development agreement within the meaning of section 1 of the National Health Service (Private Finance) Act 1997 (powers of NHS Trusts to enter into agreements).

Private finance initiative

4..–(1) A construction contract is excluded from the operation of Part II if it is a contract entered into under the private finance initiative, within the meaning given below.

(2) A contract is entered into under the private finance initiative if all the following conditions are fulfilled–
    (a) it contains a statement that it is entered into under that initiative or, as the case may be, under a project applying similar principles;
    (b) the consideration due under the contract is determined at least in part by reference to one or more of the following–
        (i) the standards attained in the performance of a service, the provision of which is the principal purpose or one of the principal purposes for which the building or structure is constructed;
        (ii) the extent, rate or intensity of use of all or any part of the building or structure in question; or
        (iii) the right to operate any facility in connection with the building or structure in question; and
    (c) one of the parties to the contract is–
        (i) a Minister of the Crown;
        (ii) a department in respect of which appropriation accounts are required to be prepared under the Exchequer and Audit Departments Act 1866;
        (iii) any other authority or body whose accounts are required to be examined and certified by or are open to the inspection of the Comptroller and Auditor General by virtue of an agreement entered into before the commencement date or by virtue of any enactment;
        (iv) any authority or body listed in Schedule 4 to the National Audit Act 1983 (nationalised industries and other public authorities);
        (v) a body whose accounts are subject to audit by auditors appointed by the Audit Commission;
        (vi) the governing body or trustees of a voluntary school within the meaning of section 31 of the Education Act 1996 (county schools and voluntary schools), or
        (vii) a company wholly owned by any of the bodies described in paragraphs (i) to (v).

Finance agreements

5.-(1) A construction contract is excluded from the operation of Part II if it is a finance agreement, within the meaning given below.

(2) A contract is a finance agreement if it is any one of the following–
- (a) any contract of insurance;
- (b) any contract under which the principal obligations include the formation or dissolution of a company, unincorporated association or partnership;
- (c) any contract under which the principal obligations include the creation or transfer of securities or any right or interest in securities;
- (d) any contract under which the principal obligations include the lending of money;
- (e) any contract under which the principal obligations include an undertaking by a person to be responsible as surety for the debt or default of another person, including a fidelity bond, advance payment bond, retention bond or performance bond.

Development agreements

6.-(1) A construction contract is excluded from the operation of Part II if it is a development agreement, within the meaning given below.

(2) A contract is a development agreement if it includes provision for the grant or disposal of a relevant interest in the land on which take place the principal construction operations to which the contract relates.

(3) In paragraph (2) above, a relevant interest in land means–
- (a) a freehold; or
- (b) a leasehold for a period which is to expire no earlier than 12 months after the completion of the construction operations under the contract.

EXPLANATORY NOTE

(This note is not part of the Order)

Part II of the Housing Grants, Construction and Regeneration Act 1996 makes provision in relation to the terms of construction contracts. Section 106 confers power on the Secretary of State to exclude descriptions of contracts from the operation of Part II. This Order excludes contracts of four descriptions.

Article 3 excludes agreements made under specified statutory provisions dealing with highway works, planning obligations, sewage works and externally financed NHS Trust agreements. Article 4 excludes agreements entered into by specified public bodies under the private finance initiative (or a project applying similar principles). Article 5 excludes agreements which primarily relate to the financing of works. Article 6 excludes development agreements, which contain provision for the disposal of an interest in land.

Appendix 3
The Scheme for Construction Contracts (England and Wales) Regulations 1998

APPENDIX 3
THE SCHEME FOR CONSTRUCTION CONTRACTS (ENGLAND AND WALES) REGULATIONS 1998

Statutory Instrument 1998 No. 649

The Secretary of State, in exercise of the powers conferred on him by sections 108(6), 114 and 146(1) and (2) of the Housing Grants, Construction and Regeneration Act 1996, and of all other powers enabling him in that behalf, having consulted such persons as he thinks fit, and draft Regulations having been approved by both Houses of Parliament, hereby makes the following Regulations:

Citation, commencement, extent and interpretation

1.-(1) These Regulations may be cited as the Scheme for Construction Contracts (England and Wales) Regulations 1998 and shall come into force at the end of the period of 8 weeks beginning with the day on which they are made (the "commencement date").

(2) These Regulations shall extend only to England and Wales.

(3) In these Regulations, "the Act" means the Housing Grants, Construction and Regeneration Act 1996.

The Scheme for Construction Contracts

2. Where a construction contract does not comply with the requirements of section 108(1) to (4) of the Act, the adjudication provisions in Part I of the Schedule to these Regulations shall apply.

3. Where–
    (a) the parties to a construction contract are unable to reach agreement for the purposes mentioned respectively in sections 109, 111 and 113 of the Act, or
    (b) a construction contract does not make provision as required by section 110 of the Act,

179

the relevant provisions in Part II of the Schedule to these Regulations shall apply.

4. The provisions in the Schedule to these Regulations shall be the Scheme for Construction Contracts for the purposes of section 114 of the Act.

SCHEDULE   Regulations 2, 3 and 4

THE SCHEME FOR CONSTRUCTION CONTRACTS
PART I–ADJUDICATION

Notice of Intention to seek Adjudication

1.–(1) Any party to a construction contract (the "referring party") may give written notice (the "notice of adjudication") of his intention to refer any dispute arising under the contract, to adjudication.

(2) The notice of adjudication shall be given to every other party to the contract.

(3) The notice of adjudication shall set out briefly–
- (a) the nature and a brief description of the dispute and of the parties involved,
- (b) details of where and when the dispute has arisen,
- (c) the nature of the redress which is sought, and
- (d) the names and addresses of the parties to the contract (including, where appropriate, the addresses which the parties have specified for the giving of notices).

2.–(1) Following the giving of a notice of adjudication and subject to any agreement between the parties to the dispute as to who shall act as adjudicator–
- (a) the referring party shall request the person (if any) specified in the contract to act as adjudicator, or
- (b) if no person is named in the contract or the person named has already indicated that he is unwilling or unable to act, and the contract provides for a specified nominating body to select a person, the referring party shall request the nominating body named in the contract to select a person to act as adjudicator, or
- (c) where neither paragraph (a) nor (b) above applies, or where the person referred to in (a) has already indicated that he is unwilling or unable to act and (b) does not apply, the referring party shall request an adjudicator nominating body to select a person to act as adjudicator.

(2) A person requested to act as adjudicator in accordance with the provisions of paragraph (1) shall indicate whether or not he is willing to act within two days of receiving the request.

(3) In this paragraph, and in paragraphs 5 and 6 below, an "adjudicator nominating body" shall mean a body (not being a natural person and not being a party to the dispute) which holds itself out publicly as a body which will select an adjudicator when requested to do so by a referring party.

3. The request referred to in paragraphs 2, 5 and 6 shall be accompanied by a copy of the notice of adjudication.

4. Any person requested or selected to act as adjudicator in accordance with paragraphs 2, 5 or 6 shall be a natural person acting in his personal capacity. A person requested or selected to act as an adjudicator shall not be an employee of any of the parties to the dispute and shall declare any interest, financial or otherwise, in any matter relating to the dispute.

5.–(1) The nominating body referred to in paragraphs 2(1)(b) and 6(1)(b) or the adjudicator nominating body referred to in paragraphs 2(1)(c), 5(2)(b) and 6(1)(c) must communicate the selection of an adjudicator to the referring party within five days of receiving a request to do so.

(2) Where the nominating body or the adjudicator nominating body fails to comply with paragraph (1), the referring party may–
    (a) agree with the other party to the dispute to request a specified person to act as adjudicator, or
    (b) request any other adjudicator nominating body to select a person to act as adjudicator.

(3) The person requested to act as adjudicator in accordance with the provisions of paragraphs (1) or (2) shall indicate whether or not he is willing to act within two days of receiving the request.

6.–(1) Where an adjudicator who is named in the contract indicates to the parties that he is unable or unwilling to act, or where he fails to respond in accordance with paragraph 2(2), the referring party may–
    (a) request another person (if any) specified in the contract to act as adjudicator, or
    (b) request the nominating body (if any) referred to in the contract to select a person to act as adjudicator, or
    (c) request any other adjudicator nominating body to select a person to act as adjudicator.

(2) The person requested to act in accordance with the provisions of paragraph (1) shall indicate whether or not he is willing to act within two days of receiving the request.

7.–(1) Where an adjudicator has been selected in accordance with paragraphs 2, 5 or 6, the referring party shall, not later than seven days from the date of the notice of adjudication, refer the dispute in writing (the "referral notice") to the adjudicator.

(2) A referral notice shall be accompanied by copies of, or relevant extracts from, the construction contract and such other documents as the referring party intends to rely upon.

(3) The referring party shall, at the same time as he sends to the adjudicator the documents referred to in paragraphs (1) and (2), send copies of those documents to every other party to the dispute.

8.–(1) The adjudicator may, with the consent of all the parties to those disputes, adjudicate at the same time on more than one dispute under the same contract.

(2) The adjudicator may, with the consent of all the parties to those disputes, adjudicate at the same time on related disputes under different contracts, whether or not one or more of those parties is a party to those disputes.

(3) All the parties in paragraphs (1) and (2) respectively may agree to extend the period within which the adjudicator may reach a decision in relation to all or any of these disputes.

(4) Where an adjudicator ceases to act because a dispute is to be adjudicated on by another person in terms of this paragraph, that adjudicator's fees and expenses shall be determined in accordance with paragraph 25.

9.–(1) An adjudicator may resign at any time on giving notice in writing to the parties to the dispute.

(2) An adjudicator must resign where the dispute is the same or substantially the same as one which has previously been referred to adjudication, and a decision has been taken in that adjudication.

(3) Where an adjudicator ceases to act under paragraph 9(1)–
   (a) the referring party may serve a fresh notice under paragraph 1 and shall request an adjudicator to act in accordance with paragraphs 2 to 7; and
   (b) if requested by the new adjudicator and insofar as it is reasonably practicable, the parties shall supply him with copies of all documents which they had made available to the previous adjudicator.

(4) Where an adjudicator resigns in the circumstances referred to in paragraph (2), or where a dispute varies significantly from the dispute referred to him in the referral notice and for that reason he is not competent to decide it, the adjudicator shall be entitled to the payment of such reasonable amount as he may determine by way of fees and expenses reasonably incurred by him. The parties shall be jointly and

severally liable for any sum which remains outstanding following the making of any determination on how the payment shall be apportioned.

10. Where any party to the dispute objects to the appointment of a particular person as adjudicator, that objection shall not invalidate the adjudicator's appointment nor any decision he may reach in accordance with paragraph 20.

11.–(1) The parties to a dispute may at any time agree to revoke the appointment of the adjudicator. The adjudicator shall be entitled to the payment of such reasonable amount as he may determine by way of fees and expenses incurred by him. The parties shall be jointly and severally liable for any sum which remains outstanding following the making of any determination on how the payment shall be apportioned.

(2) Where the revocation of the appointment of the adjudicator is due to the default or misconduct of the adjudicator, the parties shall not be liable to pay the adjudicator's fees and expenses.

Powers of the adjudicator

12. The adjudicator shall–
  (a) act impartially in carrying out his duties and shall do so in accordance with any relevant terms of the contract and shall reach his decision in accordance with the applicable law in relation to the contract; and
  (b) avoid incurring unnecessary expense.

13. The adjudicator may take the initiative in ascertaining the facts and the law necessary to determine the dispute, and shall decide on the procedure to be followed in the adjudication. In particular he may–
  (a) request any party to the contract to supply him with such documents as he may reasonably require including, if he so directs, any written statement from any party to the contract supporting or supplementing the referral notice and any other documents given under paragraph 7(2),
  (b) decide the language or languages to be used in the adjudication and whether a translation of any document is to be provided and if so by whom,
  (c) meet and question any of the parties to the contract and their representatives,
  (d) subject to obtaining any necessary consent from a third party or parties, make such site visits and inspections as he considers appropriate, whether accompanied by the parties or not,
  (e) subject to obtaining any necessary consent from a third party or parties, carry out any tests or experiments,
  (f) obtain and consider such representations and submissions as

he requires, and, provided he has notified the parties of his intention, appoint experts, assessors or legal advisers,

(g) give directions as to the timetable for the adjudication, any deadlines, or limits as to the length of written documents or oral representations to be complied with, and

(h) issue other directions relating to the conduct of the adjudication.

14. The parties shall comply with any request or direction of the adjudicator in relation to the adjudication.

15. If, without showing sufficient cause, a party fails to comply with any request, direction or timetable of the adjudicator made in accordance with his powers, fails to produce any document or written statement requested by the adjudicator, or in any other way fails to comply with a requirement under these provisions relating to the adjudication, the adjudicator may–

(a) continue the adjudication in the absence of that party or of the document or written statement requested,

(b) draw such inferences from that failure to comply as circumstances may, in the adjudicator's opinion, be justified, and

(c) make a decision on the basis of the information before him attaching such weight as he thinks fit to any evidence submitted to him outside any period he may have requested or directed.

16.–(1) Subject to any agreement between the parties to the contrary, and to the terms of paragraph (2) below, any party to the dispute may be assisted by, or represented by, such advisers or representatives (whether legally qualified or not) as he considers appropriate.

(2) Where the adjudicator is considering oral evidence or representations, a party to the dispute may not be represented by more than one person, unless the adjudicator gives directions to the contrary.

17. The adjudicator shall consider any relevant information submitted to him by any of the parties to the dispute and shall make available to them any information to be taken into account in reaching his decision.

18. The adjudicator and any party to the dispute shall not disclose to any other person any information or document provided to him in connection with the adjudication which the party supplying it has indicated is to be treated as confidential, except to the extent that it is necessary for the purposes of, or in connection with, the adjudication.

19.–(1) The adjudicator shall reach his decision not later than–

(a) twenty eight days after the date of the referral notice mentioned in paragraph 7(1), or

(b) forty two days after the date of the referral notice if the referring party so consents, or

(c) such period exceeding twenty eight days after the referral notice as the parties to the dispute may, after the giving of that notice, agree.

(2) Where the adjudicator fails, for any reason, to reach his decision in accordance with paragraph (1)

(a) any of the parties to the dispute may serve a fresh notice under paragraph 1 and shall request an adjudicator to act in accordance with paragraphs 2 to 7; and

(b) if requested by the new adjudicator and insofar as it is reasonably practicable, the parties shall supply him with copies of all documents which they had made available to the previous adjudicator.

(3) As soon as possible after he has reached a decision, the adjudicator shall deliver a copy of that decision to each of the parties to the contract.

Adjudicator's decision

20. The adjudicator shall decide the matters in dispute. He may take into account any other matters which the parties to the dispute agree should be within the scope of the adjudication or which are matters under the contract which he considers are necessarily connected with the dispute. In particular, he may–

(a) open up, revise and review any decision taken or any certificate given by any person referred to in the contract unless the contract states that the decision or certificate is final and conclusive,

(b) decide that any of the parties to the dispute is liable to make a payment under the contract (whether in sterling or some other currency) and, subject to section 111(4) of the Act, when that payment is due and the final date for payment,

(c) having regard to any term of the contract relating to the payment of interest decide the circumstances in which, and the rates at which, and the periods for which simple or compound rates of interest shall be paid.

21. In the absence of any directions by the adjudicator relating to the time for performance of his decision, the parties shall be required to comply with any decision of the adjudicator immediately on delivery of the decision to the parties in accordance with this paragraph.

22. If requested by one of the parties to the dispute, the adjudicator shall provide reasons for his decision.

Effects of the decision

23.-(1) In his decision, the adjudicator may, if he thinks fit, order any of the parties to comply peremptorily with his decision or any part of it.

(2) The decision of the adjudicator shall be binding on the parties, and they shall comply with it until the dispute is finally determined by legal proceedings, by arbitration (if the contract provides for arbitration or the parties otherwise agree to arbitration) or by agreement between the parties.

24. Section 42 of the Arbitration Act 1996 shall apply to this Scheme subject to the following modifications–
- (a) in subsection (2) for the word "tribunal" wherever it appears there shall be substituted the word "adjudicator",
- (b) in subparagraph (b) of subsection (2) for the words "arbitral proceedings" there shall be substituted the word "adjudication",
- (c) subparagraph (c) of subsection (2) shall be deleted, and
- (d) subsection (3) shall be deleted.

25. The adjudicator shall be entitled to the payment of such reasonable amount as he may determine by way of fees and expenses reasonably incurred by him. The parties shall be jointly and severally liable for any sum which remains outstanding following the making of any determination on how the payment shall be apportioned.

26. The adjudicator shall not be liable for anything done or omitted in the discharge or purported discharge of his functions as adjudicator unless the act or omission is in bad faith, and any employee or agent of the adjudicator shall be similarly protected from liability.

PART II–PAYMENT

Entitlement to and amount of stage payments

1. Where the parties to a relevant construction contract fail to agree–
- (a) the amount of any instalment or stage or periodic payment for any work under the contract, or
- (b) the intervals at which, or circumstances in which, such payments become due under that contract, or
- (c) both of the matters mentioned in sub-paragraphs (a) and (b) above,

the relevant provisions of paragraphs 2 to 4 below shall apply.

2.-(1) The amount of any payment by way of instalments or stage or periodic payments in respect of a relevant period shall be the difference between the amount determined in accordance with sub-paragraph (2) and the amount determined in accordance with sub-paragraph (3).

(2) The aggregate of the following amounts–

    (a) an amount equal to the value of any work performed in accordance with the relevant construction contract during the period from the commencement of the contract to the end of the relevant period (excluding any amount calculated in accordance with sub-paragraph (b)),

    (b) where the contract provides for payment for materials, an amount equal to the value of any materials manufactured on site or brought onto site for the purposes of the works during the period from the commencement of the contract to the end of the relevant period, and

    (c) any other amount or sum which the contract specifies shall be payable during or in respect of the period from the commencement of the contract to the end of the relevant period.

(3) The aggregate of any sums which have been paid or are due for payment by way of instalments, stage or periodic payments during the period from the commencement of the contract to the end of the relevant period.

(4) An amount calculated in accordance with this paragraph shall not exceed the difference between–

    (a) the contract price, and

    (b) the aggregate of the instalments or stage or periodic payments which have become due.

Dates for payment

3. Where the parties to a construction contract fail to provide an adequate mechanism for determining either what payments become due under the contract, or when they become due for payment, or both, the relevant provisions of paragraphs 4 to 7 shall apply.

4. Any payment of a kind mentioned in paragraph 2 above shall become due on whichever of the following dates occurs later–

    (a) the expiry of 7 days following the relevant period mentioned in paragraph 2(1) above, or

    (b) the making of a claim by the payee.

5. The final payment payable under a relevant construction contract, namely the payment of an amount equal to the difference (if any) between–

    (a) the contract price, and

    (b) the aggregate of any instalment or stage or periodic payments which have become due under the contract,

shall become due on the expiry of–

(a) 30 days following completion of the work, or

(b) the making of a claim by the payee,

whichever is the later.

6. Payment of the contract price under a construction contract (not being a relevant construction contract) shall become due on

(a) the expiry of 30 days following the completion of the work, or

(b) the making of a claim by the payee,

whichever is the later.

7. Any other payment under a construction contract shall become due

(a) on the expiry of 7 days following the completion of the work to which the payment relates, or

(b) the making of a claim by the payee,

whichever is the later.

Final date for payment

8.–(1) Where the parties to a construction contract fail to provide a final date for payment in relation to any sum which becomes due under a construction contract, the provisions of this paragraph shall apply.

(2) The final date for the making of any payment of a kind mentioned in paragraphs 2, 5, 6 or 7, shall be 17 days from the date that payment becomes due.

Notice specifying amount of payment

9. A party to a construction contract shall, not later than 5 days after the date on which any payment–

(a) becomes due from him, or

(b) would have become due, if–

(i) the other party had carried out his obligations under the contract, and

(ii) no set-off or abatement was permitted by reference to any sum claimed to be due under one or more other contracts,

give notice to the other party to the contract specifying the amount (if any) of the payment he has made or proposes to make, specifying to what the payment relates and the basis on which that amount is calculated.

Notice of intention to withhold payment

10. Any notice of intention to withhold payment mentioned in section 111 of the Act shall be given not later than the prescribed period, which is to say not later than 7 days before the final date for payment determined either in accordance with the construction contract, or where no such provision is made in the contract, in accordance with paragraph 8 above.

Prohibition of conditional payment provisions

11. Where a provision making payment under a construction contract conditional on the payer receiving payment from a third person is ineffective as mentioned in section 113 of the Act, and the parties have not agreed other terms for payment, the relevant provisions of–

(a) paragraphs 2, 4, 5, 7, 8, 9 and 10 shall apply in the case of a relevant construction contract, and

(b) paragraphs 6, 7, 8, 9 and 10 shall apply in the case of any other construction contract.

Interpretation

12. In this Part of the Scheme for Construction Contracts–

"claim by the payee" means a written notice given by the party carrying out work under a construction contract to the other party specifying the amount of any payment or payments which he considers to be due and the basis on which it is, or they are calculated;

"contract price" means the entire sum payable under the construction contract in respect of the work;

"relevant construction contract" means any construction contract other than one–

(a) which specifies that the duration of the work is to be less than 45 days, or

(b) in respect of which the parties agree that the duration of the work is estimated to be less than 45 days;

"relevant period" means a period which is specified in, or is calculated by reference to the construction contract or where no such period is so specified or is so calculable, a period of 28 days;

"value of work" means an amount determined in accordance with the construction contract under which the work is performed or where the contract contains no such provision, the cost of any work performed in accordance with that contract together with an amount equal to any overhead or profit included in the contract price;

"work" means any of the work or services mentioned in section 104 of the Act.

EXPLANATORY NOTE

(This note is not part of the Order)

Part II of the Housing Grants, Construction and Regeneration Act 1996 makes provision in relation to construction contracts. Section 114 empowers the Secretary of State to make the Scheme for Construction

Contracts. Where a construction contract does not comply with the requirements of sections 108 to 111 (adjudication of disputes and payment provisions), and section 113 (prohibition of conditional payment provisions), the relevant provisions of the Scheme for Construction Contracts have effect.

The Scheme which is contained in the Schedule to these Regulations is in two parts. Part I provides for the selection and appointment of an adjudicator, gives powers to the adjudicator to gather and consider information, and makes provisions in respect of his decisions. Part II makes provision with respect to payments under a construction contract where either the contract fails to make provision or the parties fail to agree–

(a) the method for calculating the amount of any instalment, stage or periodic payment,

(b) the due date and the final date for payments to be made, and

(c) prescribes the period within which a notice of intention to withhold payment must be given.

Appendix 4
Contract Adjudication Provisions and Related Procedures

ICE CONDITIONS OF CONTRACT, 7TH EDITION
(Extract)
ICE ADJUDICATION PROCEDURE (1997)

APPENDIX 4.1
ICE CONDITIONS OF CONTRACT 7TH EDITION (Extract)

AVOIDANCE AND SETTLEMENT OF DISPUTES

Avoidance of disputes 66

(1)   In order to overcome where possible the causes of disputes and in those cases where disputes are likely still to arise to facilitate their clear definition and early resolution (whether by agreement or otherwise) the following procedure shall apply for the avoidance and settlement of disputes.

Matters of dissatisfaction

(2)   If at any time

(a)   the Contractor is dissatisfied with any act or instruction of the Engineer's Representative or any other person responsible to the Engineer or

(b)   the Employer or the Contractor is dissatisfied with any decision opinion instruction direction certificate or valuation of the Engineer or with any other matter arising under or in connection with the Contract or the carrying out of the Works

the matter of dissatisfaction shall be referred to the Engineer who shall notify his written decision to the Employer and the Contractor within one month of the reference to him.

Disputes

(3)   The Employer and the Contractor agree that no matter shall constitute nor be said to give rise to a dispute unless and until in respect of that matter

(a)   the time for the giving of a decision by the Engineer on a matter of dissatisfaction under Clause 66(2) has expired or the decision given is unacceptable or has not been implemented and in consequence the Employer or the Contractor has served on the other and on the Engineer a notice in writing (hereinafter called the Notice of Dispute) or

(b)   an adjudicator has given a decision on a dispute under Clause 66(6) and the Employer or the Contractor is not giving effect to the decision, and in consequence the other has served on him and the Engineer a Notice of Dispute and the dispute shall be that stated in the Notice of Dispute. For the purposes of all matters arising under or in connection with the Contract or the carrying out of the Works the word "dispute" shall be construed accordingly and shall include any difference.

(4) (a) Notwithstanding the existence of a dispute following the service of a Notice under Clause 66(3) and unless the Contract has already been determined or abandoned the Employer and the Contractor shall continue to perform their obligations.

(b) The Employer and the Contractor shall give effect forthwith to every decision of

> (i) the Engineer on a matter of dissatisfaction given under Clause 66(2) and

> (ii) the adjudicator on a dispute given under Clause 66(6)

unless and until that decision is revised by agreement of the Employer and Contractor or pursuant to Clause 66.

. . .

Adjudication (6) (a) The Employer and the Contractor each has the right to refer a dispute as to a matter under the Contract for adjudication and either party may give notice in writing (hereinafter called the Notice of Adjudication) to the other at any time of his intention so to do. The adjudication shall be conducted under "The Institution of Civil Engineers' Adjudication Procedure 1997" or any amendment or modification thereof being in force at the time of the said Notice.

(b) Unless the adjudicator has already been appointed he is to be appointed by a timetable with the object of securing his appointment and referral of the dispute to him within 7 days of such notice.

(c) The adjudicator shall reach a decision within 28 days of referral or such longer period as is agreed by the parties after the dispute has been referred.

(d) The adjudicator may extend the period of 28 days by up to 14 days with the consent of the party by whom the dispute was referred.

(e) The adjudicator shall act impartially.

(f) The adjudicator may take the initiative in ascertaining the facts and the law.

(7) The decision of the adjudicator shall be binding until the dispute is finally determined by legal proceedings or by arbitration (if the contract provides for arbitration or the parties otherwise agree to arbitration) or by agreement.

(8) The adjudicator is not liable for anything done or omitted in the discharge or purported discharge of his functions as adjudicator unless the act or omission is in bad faith and any employee or agent of the adjudicator is similarly not liable.

Arbitration

(9) (a) All disputes arising under or in connection with the Contract or the carrying out of the Works other than failure to give effect to a decision of an adjudicator shall be finally determined by reference to arbitration. The party seeking arbitration shall serve on the other party a notice in writing (called the Notice to Refer) to refer the dispute to arbitration.

(b) Where an adjudicator has given a decision under Clause 66(6) in respect of the particular dispute the Notice to Refer must be served within three months of the giving of the decision otherwise it shall be final as well as binding.

ICE ADJUDICATION PROCEDURE (1997)

1. General principles

1.1 The adjudication shall be conducted in accordance with the edition of the ICE Adjudication Procedure which is current at the date of issue of a notice in writing of intention to refer a dispute to adjudication (hereinafter called the Notice of Adjudication) and the Adjudicator shall be appointed under the Adjudicator's Agreement which forms a part of this Procedure. If a conflict arises between this Procedure and the Contract then this Procedure shall prevail.

1.2 The object of adjudication is to reach a fair, rapid and inexpensive determination of a dispute arising under the Contract and this Procedure shall be interpreted accordingly.

1.3 The Adjudicator shall be a named individual and shall act impartially.

1.4 In making a decision, the Adjudicator may take the initiative in ascertaining the facts and the law. The adjudication shall be neither an expert determination nor an arbitration but the Adjudicator may rely on his own expert knowledge and experience.

1.5 The Adjudicator's decision shall be binding until the dispute is finally determined by legal proceedings, by arbitration (if the contract provides for arbitration or the Parties otherwise agree to arbitration) or by agreement.

1.6 The Parties shall implement the Adjudicator's decision without delay whether or not the dispute is to be referred to legal proceedings or arbitration. Payment shall be made in accordance with the payment provisions in the Contract, in the next stage payment which becomes due after the date of issue of the decision, unless otherwise directed by the Adjudicator or unless the decision is in relation to an effective notice under Section 111(4) of the Act.

2. The Notice of Adjudication

2.1 Any Party may give notice at any time of its intention to refer a dispute arising under the Contract to adjudication by giving a written Notice of Adjudication to the other Party. The Notice of Adjudication shall include:

(a) the details and date of the Contract between the Parties;
(b) the issues which the Adjudicator is being asked to decide;
(c) details of the nature and extent of the redress sought.

3. The appointment of the Adjudicator

3.1 Where an Adjudicator has either been named in the Contract or agreed by the Parties prior to the issue of the Notice of Adjudication the Party issuing the Notice of Adjudication shall at the same time send to the Adjudicator a copy of the Notice of Adjudication and a request for confirmation, within four days of the date of issue of the Notice of Adjudication, that the Adjudicator is able and willing to act.

3.2 Where an Adjudicator has not been so named or agreed the Party issuing the Notice of Adjudication may include with the Notice the names of one or more persons with their addresses who have agreed to act, any one of whom would be acceptable to the referring Party, for selection by the other Party. The other Party shall select and notify the referring Party and the selected Adjudicator within four days of the date of issue of the Notice of Adjudication.

3.3 If confirmation is not received under paragraph 3.1 or a selection is not made under paragraph 3.2 or the Adjudicator does not accept or is unable to act then either Party may within a further three days request the person or body named in the Contract or if none is so named The Institution of Civil Engineers to appoint the Adjudicator. Such request shall be in writing or the appropriate form of application for the appointment of an adjudicator and accompanied by a copy of the Notice of Adjudication and the appropriate fee.

3.4 The Adjudicator shall be appointed on the terms and conditions set out in the attached Adjudicator's Agreement and Schedule and shall be entitled to be paid a reasonable fee together with his expenses. The Parties shall sign the agreement within 7 days of being requested to do so.

3.5 If for any reason whatsoever the Adjudicator is unable to act, either Party may require the appointment of a replacement adjudicator in accordance with the procedure in paragraph 3.3.

4. Referral

4.1 The referring Party shall within two days of receipt of confirmation under 3.1, or notification of selection under 3.2, or appointment under 3.3 send to the Adjudicator, with a copy to the other Party, a full statement of his case which should include:

(a) a copy of the Notice of Adjudication;
(b) a copy of any adjudication provision in the Contract, and
(c) the information upon which he relies, including supporting documents.

4.2　The date of referral of the dispute to adjudication shall be the date upon which the Adjudicator receives the documents referred to in paragraph 4.1. The Adjudicator shall notify the Parties forthwith of that date.

5. Conduct of the adjudication

5.1　The Adjudicator shall reach his decision within 28 days of referral, or such longer period as is agreed by the Parties after the dispute has been referred. The period of 28 days may be extended by up to 14 days with the consent of the referring party.

5.2　The Adjudicator shall determine the matters set out in the Notice of Adjudication, together with any other matters which the Parties and the Adjudicator agree should be within the scope of the adjudication.

5.3　The Adjudicator may open up review and revise any decision, (other than that of an adjudicator unless agreed by the Parties), opinion, instruction, direction, certificate or valuation made under or in connection with the Contract and which is relevant to the dispute. He may order the payment of a sum of money, or other redress but no decision of the Adjudicator shall affect the freedom of the Parties to vary the terms of the Contract or the Engineer or other authorised person to vary the Works in accordance with the Contract.

5.4　The other Party may submit his response to the statement under paragraph 4.1 within 14 days of referral. The period of response may be extended by agreement between the Parties and the Adjudicator.

5.5　The Adjudicator shall have complete discretion as to how to conduct the adjudication, and shall establish the procedure and timetable, subject to any limitation that there may be in the Contract or the Act. He shall not be required to observe any rule of evidence, procedure or otherwise, of any court. Without prejudice to the generality of these powers, he may:

    (a) ask for further written information;
    (b) meet and question the Parties and their representatives;
    (c) visit the site;
    (d) request the production of documents or the attendance of people whom he considers could assist;
    (e) set times for (a) – (d) and similar activities;
    (f) proceed with the adjudication and reach a decision even if a Party fails:
        (i)　to provide information;
        (ii)　to attend a meeting;
        (iii) to take any other action requested by the Adjudicator;

(g) issue such further directions as he considers to be appropriate.

5.6   The Adjudicator may obtain legal or technical advice having first notified the Parties of his intention.

5.7   Any Party may at any time ask that additional Parties shall be joined in the Adjudication. Joinder of additional Parties shall be subject to the agreement of the Adjudicator and the existing and additional Parties. An additional Party shall have the same rights and obligations as the other Parties, unless otherwise agreed by the Adjudicator and the Parties.

6. The Decision

6.1   The Adjudicator shall reach his decision and so notify the Parties within the time limits in paragraph 5.1 and may reach a decision on different aspects of the dispute at different times. He shall not be required to give reasons.

6.2   The Adjudicator may in any decision direct the payment of such simple or compound interest at such rate and between such dates or events as he considers appropriate.

6.3   Should the Adjudicator fail to reach his decision and notify the Parties in the due time either Party may give seven days notice of its intention to refer the dispute to a replacement adjudicator appointed in accordance with the procedures in paragraph 3.3.

6.4   If the Adjudicator fails to reach and notify his decision in due time but does so before the dispute has been referred to a replacement adjudicator under paragraph 6.3 his decision shall still be effective.

If the Parties are not so notified then the decision shall be of no effect and the Adjudicator shall not be entitled to any fees or expenses but the Parties shall be responsible for the fees and expenses of any legal or technnical adviser appointed under paragraph 5.6 subject to the Parties having received such advice.

6.5   The Parties shall bear their own costs and expenses incurred in the adjudication. The Parties shall be jointly and severally responsible for the Adjudicator's fees and expenses, including those of any legal or technical adviser appointed under paragraph 5.6, but in his decision the Adjudicator may direct a Party to pay all or part of his fees and expenses. If he makes no such direction the Parties shall pay them in equal shares.

6.6   At any time until 7 days before the Adjudicator is due to reach his decision, he may give notice to the Parties that he will deliver it only on full payment of his fees and expenses. Any Party may then pay these costs in order to obtain the decision and recover the other Party's share of the costs in accordance with paragraph 6.5 as a debt due.

6.7 The Parties shall be entitled to the relief and remedies set out in the decision and to seek summary enforcement thereof, regardless of whether the dispute is to be referred to legal proceedings or arbitration. No issue decided by an adjudicator may subsequently be laid before another adjudicator unless so agreed by the Parties.

6.8 In the event that the dispute is referred to legal proceedings or arbitration, the Adjudicator's decision shall not inhibit the court or arbitrator from determining the Parties' rights or obligations anew.

6.9 The Adjudicator may on his own initiative, or at the request of either Party, correct a decision so as to remove any clerical mistake, error or ambiguity provided that the initiative is taken, or the request is made within 14 days of the notification of the decision to the Parties. The Adjudicator shall make his corrections within 7 days of any request by a Party.

7. Miscellaneous provisions

7.1 Unless the Parties agree, the Adjudicator shall not be appointed arbitrator in any subsequent arbitration between the Parties under the Contract. No Party may call the Adjudicator as a witness in any legal proceedings or arbitration concerning the subject matter of the adjudication.

7.2 The Adjudicator shall not be liable for anything done or omitted in the discharge or purported discharge of his functions as Adjudicator unless the act or omission is in bad faith, and any employee or agent of the Adjudicator shall be similarly protected from liability. The Parties shall save harmless and indemnify the Adjudicator and any employee or agent of the Adjudicator against all claims by third parties and in respect of this shall be jointly and severally liable.

7.3 Neither The Institution of Civil Engineers nor its servants or agents shall be liable to any Party for any act omission or misconduct in connection with any appointment made or any adjudication conducted under this Procedure.

7.4 All notices shall be sent by recorded delivery to the address stated in the Contract for service of notices, or if none, the principal place of business or registered office (in the case of a company). Any agreement required by this Procedure shall be evidenced in writing.

7.5 This Procedure shall be interpreted in accordance with the law of the Contract.

Appendix 4

8. Definitions   8.1   (a) The "Act" means the Housing Grants, Construction and Regeneration Act 1996.

(b) The "Adjudicator" means the person named as such in the Contract or appointed in accordance with this Procedure.

(c) "Contract" means the the contract or the agreement between the Parties which contains the provision for adjudication.

(d) "Party" means a Party to the Contract and references to either Party or the other Party or Parties shall include any additional Party or Parties joined in accordance with this Procedure.

9. Application to particular contracts   9.1   When this Procedure is used with The Institution of Civil Engineers' Agreement for Consultancy Work in Respect of Domestic or Small Works the Adjudicator may determine any dispute in connection with or arising out of the Contract.

Appendix 4.2
JCT Standard Form of Building Contract 1998 Amendment 4, 2002 (Extract)

APPENDIX 4.2
JCT STANDARD FORM OF BUILDING CONTRACT, 1998 AMENDMENT 4, 2002 (Extract)

Part 4: Settlement of disputes – adjudication – arbitration – legal proceedings [uu]

41A   Adjudication [uu·1]

Application of clause 41A

41A·1 Clause 41A applies where, pursuant to article 5, either Party refers any dispute or difference arising under this Contract to adjudication.

Identity of Adjudicator

41A·2 The Adjudicator to decide the dispute or differences shall be either an individual agreed by the Parties or, on the application of either Party, an individual to be nominated as the Adjudicator by the person named in the Appendix ('the nominator'). Provided that [vv]

41A·2 ·1 no Adjudicator shall be agreed or nominated under clause 41A·2 or clause 41A·3 who will not execute the Standard Agreement for the appointment of an Adjudicator issued by the JCT (the 'JCT Adjudication Agreement' [ww]) with the Parties, [vv] and

41A·2 ·2 where either Party has given notice of his intention to refer a dispute or difference to adjudication then

– any agreement by the Parties on the appointment of an adjudicator must be reached with the object of securing the appointment of, and the referral of the dispute or difference to, the Adjudicator within 7 days of the date of the notice of intention to refer (*see clause 41A·4·1*);

– any application to the nominator must be made with the object of securing the appointment of, and the referral of the dispute or difference to, the Adjudicator within 7 days of the date of the notice of intention to refer.

Upon agreement by the Parties on the appointment of the Adjudicator or upon receipt by the Parties from the nominator of the name of the nominated Adjudicator the Parties shall thereupon execute with the Adjudicator the JCT Adjudication Agreement.

Death of Adjudicator – inability to adjudicate

41A·3 If the Adjudicator dies or becomes ill or is unavailable for some other cause and is thus unable to adjudicate on a dispute or difference referred to him, then either the Parties may agree upon an individual to replace the Adjudicator or either Party may apply to the nominator for the nomination of an adjudicator to adjudicate that dispute or difference; and the Parties shall execute the JCT Adjudication Agreement with the agreed or nominated Adjudicator.

Dispute or difference – notice of intention to refer to adjudication – referral

41A·4 .1 When pursuant to article 5 a Party requires a dispute or difference to be referred to adjudication then that Party shall give notice to the other Party of his intention to refer the dispute or difference, briefly identified in the notice, to adjudication. If an Adjudicator is agreed or appointed within 7 days of the notice then the Party giving the notice shall refer the dispute or difference to the Adjudicator ('the referral') within 7 days of the notice. If an Adjudicator is not agreed or appointed within 7 days of the notice the referral shall be made immediately on such agreement or appointment. The said Party shall include with that referral particulars of the dispute or difference together with a summary of the contentions on which he relies, a statement of the relief or remedy which is sought and any material he wishes the Adjudicator to consider. The referral and its accompanying documentation shall be copied simultaneously to the other Party.

Footnotes

[uu] It is open to the Employer and the Contractor to resolve disputes by the process of Mediation: see Practice Note 28 'Mediation on a Building Contractor or Sub-Contract Dispute'.

[uu·1] The time periods generally specified in this clause are those defined by statute. Where the nature of the dispute or the work concerned may have any significant effect upon the progress or cost of the Works such as works relating to the primary structural elements the Adjudicator should consider an accelerated time table for the adjudication procedures: see JCT Practice Note 2 (Series 2): Adjudication under JCT Forms.

[vv] The nominators named in the Appendix have agreed with the JCT that they will comply with the requirements of clause 41A on the nomination of an adjudicator including the requirement in clause **41A·2·2** for the nomination to be made with the object of securing the appointment of, and the referral of the dispute or difference to, the Adjudicator within 7 days of the date of the notice of intention to refer; and will only nominate adjudicators who will enter into the 'JCT Adjudication Agreement'.

[ww] The JCT Adjudication Agreement is available from the retailers of JCT Forms. A version of this Agreement is also available for use if the Parties have named an Adjudicator in their contract.

41A·4 ·2   The referral by a Party with its accompanying documentation to the Adjudicator and the copies thereof to be provided to the other Party shall be given by actual delivery or by FAX or by special delivery or recorded delivery. If given by FAX then, for record purposes, the referral and its accompanying documentation must forthwith be sent by first class post or given by actual delivery. If sent by special delivery or recorded delivery the referral and its accompanying documentation shall, subject to proof to the contrary, be deemed to have been received 48 hours after the date of posting subject to the exclusion of Sundays and any Public Holiday.

Conduct of the adjudication   **41A·5** .1   The Adjudicator shall immediately upon receipt of the referral and its accompanying documentation confirm the date of that receipt to the Parties.

41A.5 .2   The Party not making the referral may, by the same means stated in clause 41A·4·2, send to the Adjudicator within 7 days of the date of the referral, with a copy to the other Party, a written statement of the contentions on which he relies and any material he wishes the Adjudicator to consider.

41A·5 .3   The Adjudicator shall within 28 days of the referral under clause 41A·4·1 and acting as an Adjudicator for the purposes of S.108 of the Housing Grants, Construction and Regeneration Act 1996 and not as an expert or an arbitrator reach his decision and forthwith send that decision in writing to the Parties. Provided that the Party who has made the referral may consent to allowing the Adjudicator to extend the period of 28 days by up to 14 days; and that by agreement between the Parties after the referral has been made a longer period than 28 days may be notified jointly by the Parties to the Adjudicator within which to reach his decision.

41A·5 ·4   The Adjudicator shall not be obliged to give reasons for his decision.

41A·5 .5   In reaching his decision the Adjudicator shall act impartially and set his own procedure; and at his absolute discretion may take the initiative in ascertaining the facts and the law as he considers necessary in respect of the referral which may include the following:

  ·5 ·1   using his own knowledge and/or experience;

  ·5 ·2   subject to clause 30·9, opening up, reviewing and revising any certificate, opinion, decision, requirement or notice issued, given or made under this Contract as if no such certificate, opinion, decision, requirement or notice had been issued, given or made;

·5 ·3 requiring from the Parties further information than that contained in the notice of referral and its accompanying documentation or in any written statement provided by the Parties including the results of any tests that have been made or of any opening up;

·5 ·4 requiring the Parties to carry out tests or additional tests or to open up work or further open up work;

·5 ·5 visiting the site of the Works or any workshop where work is being or has been prepared for this Contract;

·5 ·6 obtaining such information as he considers necessary from any employee or representative of the Parties provided that before obtaining information from an employee of a Party he has given prior notice to that Party;

·5 ·7 obtaining from others such information and advice as he considers necessary on technical and on legal matters subject to giving prior notice to the Parties together with a statement or estimate of the cost involved:

·5 ·8 having regard to any term of this Contract relating to the payment of interest, deciding the circumstances in which or the period for which a simple rate of interest shall be paid.

41A·5 ·6 Any failure by either Party to enter into the JCT Adjudication Agreement or to comply with any requirement of the Adjudicator under clause 41A·5·5 or with any provision in or requirement under clause 41A shall not invalidate the decision of the Adjudicator.

41A·5 ·7 The Parties shall meet their own costs of the adjudication except that the Adjudicator may direct as to who should pay the cost of any test or opening up if required pursuant to clause 41A·5·5·4.

41A·5 ·8 Where any dispute or difference arises under clause 8·4·4 as to whether an instruction issued thereunder is reasonable in all the circumstances the following provisions shall apply:

·8 ·1 The Adjudicator to decide such dispute or difference shall (where practicable) be an individual with appropriate expertise and experience in the specialist area or discipline relevant to the instruction or issue in dispute.

·8 ·2 Where the Adjudicator does not have the appropriate expertise and experience referred to in clause 41A·5·8·1 above the Adjudicator shall appoint an independent expert with such relevant expertise and experience to advise and report in writing on whether or not any instruction issued under clause 8·4·4 is reasonable in all the circumstances.

·8 ·3 Where an expert has been appointed by the Adjudicator pursuant to clause 41A·5·8·2 above the Parties shall be jointly and severally responsible for the expert's fees and expenses but, in his decision, the Adjudicator shall direct as to who should pay the fees and expenses of such expert or the proportion in which such fees and expenses are to be shared between the Parties.

·8 ·4 Notwithstanding the provisions of clause 41A·5·4 above, where an independent expert has been appointed by the Adjudicator pursuant to clause 41A·5·8·2 above, copies of the Adjudicator's instructions to the expert and any written advice or reports received from such expert shall be supplied to the Parties as soon as practicable.

Adjudicator's fee and reasonable expenses – payment

41A·6 ·1 The Adjudicator in his decision shall state how payment of his fee and reasonable expenses is to be apportioned as between the Parties. In default of such statement the Parties shall bear the cost of the Adjudicator's fee and reasonable expenses in equal proportions.

41A·6 ·2 The Parties shall be jointly and severally liable to the Adjudicator for his fee and for all expenses reasonably incurred by the Adjudicator pursuant to the adjudication.

Effect of Adjudicator's decision

41A·7 ·1 The decision of the Adjudicator shall be binding on the Parties until the dispute or difference is finally determined by arbitration or by legal proceedings [xx] or by an agreement in writing between the Parties made after the decision of the Adjudicator has been given.

41A·7 ·2 The Parties shall, without prejudice to their other rights under this Contract, comply with the decision of the Adjudicator; and the Employer and the Contractor shall ensure that the decision of the Adjudicator is given effect.

41A·7 ·3 If either Party does not comply with the decision of the Adjudicator the other Party shall be entitled to take legal proceedings to secure such compliance pending any final determination of the referred dispute or difference pursuant to clause 41A·7·1.

Immunity

41A·8 The Adjudicator shall not be liable for anything done or omitted in the discharge or purported discharge of his functions as

211

Adjudicator unless the act or omission is in bad faith and this protection from liability shall similarly extend to any employee or agent of the Adjudicator.

Footnote   [xx] The arbitration or legal proceedings are *not* an appeal against the decision of the Adjudicator but are a consideration of the dispute or difference as if no decision had been made by an Adjudicator.

Appendix 4.3
Engineering and Construction Contract, Second Edition (Extract) Option Y(UK)2 1998 (Extract)

APPENDIX 4.3
ENGINEERING AND CONSTRUCTION CONTRACT, SECOND EDITION (Extract)

9 Disputes and termination

Settlement of
disputes

90

90.1 Any dispute arising under or in connection with this contract is submitted to and settled by the *Adjudicator* as follows.

ADJUDICATION TABLE

Dispute about:	Which Party may submit it to the *Adjudicator*?	When may it be submitted to the *Adjudicator*?
An action of the *Project Manager* or the *Supervisor*	The *Contractor*	Between two and four weeks after the *Contractor's* notification of the dispute to the *Project Manager*, the notification itself being made not more than four weeks after the *Contractor* becomes aware of the action
The *Project Manager* or *Supervisor* not having taken an action	The *Contractor*	Between two and four weeks after the *Contractor's* notification of the dispute to the *Project Manager*, the notification itself being made not more than four weeks after the *Contractor* becomes aware that the action was not taken
Any other matter	Either Party	Between two and four weeks after notification of the dispute to the other Party and the *Project Manager*

90.2   The *Adjudicator* settles the dispute by notifying the Parties and the *Project Manager* of his decision together with his reasons within the time allowed by this contract. Unless and until there is such a settlement, the Parties and the *Project Manager* proceed as if the action, inaction or other matter disputed were not disputed. The decision is final and binding unless and until revised by the *tribunal*.

The adjudication   **91**

91.1   The Party submitting the dispute to the *Adjudicator* includes with his submission information to be considered by the *Adjudicator*. Any further information from a Party to be considered by the *Adjudicator* is provided within four weeks from the submission. The *Adjudicator* notifies his decision within four weeks of the end of the period for providing information. The four week periods in this clause may be extended if requested by the *Adjudicator* in view of the nature of the dispute and agreed by the Parties.

91.2   If a matter disputed under or in connection with a subcontract is also a matter disputed under or in connection with this contract, the *Contractor* may submit the subcontract dispute to the *Adjudicator* at the same time as the main contract submission. The *Adjudicator* then settles the two disputes together and references to the Parties for the purposes of the dispute are interpreted as including the Subcontractor.

The Adjudicator   **92**

92.1   The *Adjudicator* settles the dispute as independent adjudicator and not as arbitrator. His decision is enforceable as a matter of contractual obligation between the Parties and not as an arbitral award. The *Adjudicator*'s powers include the power to review and revise any action or inaction of the *Project Manager or Supervisor* related to the dispute. Any communication between a Party and the *Adjudicator* is communicated also to the other Party. If the *Adjudicator*'s decision includes assessment of additional cost or delay caused to the *Contractor*, he makes his assessment in the same way as a compensation event is assessed.

92.2   If the *Adjudicator* resigns or is unable to act, the Parties choose a new adjudicator jointly. If the Parties have not chosen a new adjudicator jointly within four weeks of the *Adjudicator* resigning or becoming unable to act, a Party may ask the person stated in the Contract Data to choose a new adjudicator and the Parties accept his choice. The new adjudicator is appointed as *Adjudicator* under the NEC Adjudicator's Contract. He has power to settle disputes that were currently submitted to his predecessor but had not been settled at the time when his predecessor resigned or became unable to act. The date of his appointment is the date of submission of these disputes to him as *Adjudicator*.

Review by the tribunal   **93**

93.1   If after the *Adjudicator*

- notifies his decision or
- fails to do so

within the time provided by this contract a Party is dissatisfied, that Party notifies the other Party of his intention to refer the matter which he disputes to the *tribunal*. It is not referable to the *tribunal* unless the dissatisfied Party notifies his intention within four weeks of

- notification of the *Adjudicator*'s decision or
- the time provided by this contract for this notification if the *Adjudicator* fails to notify his decision within that time

whichever is the earlier. The *tribunal* proceedings are not started before Completion of the whole of the *works* or earlier termination.

93.2   The *tribunal* settles the dispute referred to it. Its powers include the power to review and revise any decision of the *Adjudicator* and any action or inaction of the *Project Manager* or the *Supervisor* related to the dispute. A Party is not limited in the *tribunal* proceedings to the information, evidence or arguments put to the *Adjudicator*.

OPTION Y(UK)2 1998 (Extract)

Option Y(UK)2: The Housing Grants, Construction and Regeneration Act 1996

Y2.5

Clause 90 is deleted and replaced by the following:

Avoidance and settlement of disputes

90

90.1 The Parties and the *Project Manager* follow this procedure for the avoidance and settlement of disputes.

90.2 If the *Contractor* is dissatisfied with an action or a failure to take action by the *Project Manager*, he notifies his dissatisfaction to the *Project Manager* no later than

- four weeks after he became aware of the action or
- four weeks after he became aware that the action had not been taken.

Within two weeks of such notification of dissatisfaction, the *Contractor* and the *Project Manager* attend a meeting to discuss and seek to resolve the matter.

90.3 If either Party is dissatisfied with any other matter, he notifies his dissatisfaction to the *Project Manager* and to the other Party no later than four weeks after he became aware of the matter. Within two weeks of such notification of dissatisfaction, the Parties and the *Project Manager* attend a meeting to discuss and seek to resolve the matter.

90.4 The Parties agree that no matter shall be a dispute unless a notice of dissatisfaction has been given and the matter has not been resolved within four weeks. The word dispute (which includes a difference) has that meaning.

90.5 Either Party may give notice to the other Party at any time of his intention to refer a dispute to adjudication. The notifying Party refers the dispute to the *Adjudicator* within seven days of the notice.

90.6 The Party referring the dispute to the *Adjudicator* includes with his submission information to be considered by the *Adjudicator*. Any further information from a Party to be considered by the *Adjudicator* is provided within fourteen days of referral.

90.7 Unless and until the *Adjudicator* has given his decision on the dispute, the Parties and the *Project Manager* proceed as if the action, failure to take action or other matters were not disputed.

90.8 The *Adjudicator* acts impartially. The *Adjudicator* may take the initiative in ascertaining the facts and the law.

90.9 The *Adjudicator* reaches a decision within twenty eight days of referral or such longer period as is agreed by the Parties after the dispute has been referred. The *Adjudicator* may extend the period of twenty eight days by up to fourteen days with the consent of the notifying Party.

90.10 The *Adjudicator* provides his reasons to the Parties and to the *Project Manager* with his decision.

90.11 The decision of the *Adjudicator* is binding until the dispute is finally determined by the *tribunal* or by agreement.

90.12 The *Adjudicator* is not liable for anything done or omitted in the discharge or purported discharge of his functions as adjudicator unless the act or omission is in bad faith and any employee or agent of the *Adjudicator* is similarly protected from liability.

Y2.6 **Clause 91 is amended as follows:-**

Side heading **"The adjudication"** is replaced with **"Combining procedures"**

Clause 91.1 is deleted and replaced by the following:-

91.1 If a matter causing dissatisfaction under or in connection with a subcontract is also a matter causing dissatisfaction under or in connection with this contract, the subcontractor may attend the meeting between the Parties and the *Project Manager* to discuss and seek to resolve the matter.

Clause 91.2 line 4 "settles" is replaced with "gives his decision on"

Y2.7 **Clause 92 is amended as follows:-**

Clause 92.1 line 1 "settles" is replaced with "gives his decision on"

Clause 92.2 line 6 "settle" is replaced with "decide on"

Clause 92.2 line 7 "had not been settled" is replaced with "a decision had not been given"

Appendix 5
Independent Adjudication Rules and Procedures

Appendix 5.1
Technology and Construction Solicitors Association 2002 Version 2.0 Procedural Rules for Adjudication

TECHNOLOGY AND CONSTRUCTION SOLICITORS ASSOCIATION 2002 Version 2.0 Procedural Rules for Adjudication

1.   The following rules

(i)   may be incorporated into any contract by reference to the "TeCSA Adjudication Rules" or the "ORSA Adjudication Rules", which expressions shall mean, in relation to any adjudication, the most recent edition hereof as at the date of the written notice requiring that adjudication.

(ii)   meet the requirements of adjudication procedure as set out in section 108 of the Housing Grants, Construction and Regeneration Act 1996; Part I of the Scheme for Construction Contracts shall thus not apply.

DEFINITIONS

2.   In these Rules:

"Contract"   means the agreement which includes the agreement to adjudicate in accordance with these Rules

"Party"   means any party to the Contract

"Chairman"   means the Chairman for the time being of the Technology and Construction Solicitors Association ("TeCSA"), or such other officer thereof as is authorised to deputise for him.

"days"   shall have the same meaning as and be calculated in accordance with Part II of the Housing Grants, Construction and Regeneration Act 1996.

COMMENCEMENT AND APPOINTMENT

3.   These Rules shall apply upon any Party giving written notice to any other Party requiring adjudication, and identifying in general terms the dispute in respect of which adjudication is required.

4.   Where the Parties have agreed upon the identity of an adjudicator who confirms his readiness and willingness to embark upon the Adjudication within 7 days of the notice requiring adjudication, then that person shall be the Adjudicator.

5.   Where the Parties have not so agreed upon an adjudicator, or where such person has not so confirmed his willingness to act, then any Party shall apply to the Chairman of TeCSA for a nomination. The following procedure shall apply:-

(i)   The application shall be in writing, accompanied by a copy of the Contract, a copy of the written notice requiring adjudication, and TeCSA's appointment fee of £100.

(ii)   The Chairman of TeCSA shall endeavour to secure the appointment of an Adjudicator within 7 days from the notice requiring adjudication.

(iii)   Any person so appointed, and not any person named in the Contract whose readiness or willingness is in question, shall be the Adjudicator.

6.   Within 7 days from the date of the Notice referred to in Rule 3:-

(i)   provided he is willing and able to act, any agreed Adjudicator under Rule 4 or nominated Adjudicator under Rule 5(ii) shall give written notice of his acceptance of appointment to all parties; and

(ii)   the referring party shall serve the Referral Notice on the Adjudicator and the Responding Party.

7.   The date of the referral of the dispute shall be the date the Referral Notice is received by the Adjudicator. The Adjudicator shall confirm to the Parties the date of receipt of the Referral Notice.

8.   The Chairman of TeCSA shall have the power by written notice to the Parties to replace the Adjudicator with another nominated person if and when it appears necessary to him to do so. The Chairman of TeCSA shall consider whether to exercise such power if any Party shall represent to him that the Adjudicator is not acting impartially, or that the Adjudicator is physically or mentally incapable of conducting the Adjudication, or that the Adjudicator is failing with necessary dispatch to proceed with the Adjudication or make his decision. In the event of a replacement under this Rule, directions and decisions of the previous Adjudicator shall remain in effect unless reviewed and replaced by the new Adjudicator, and all timescales shall be recalculated from the date of the replacement. Any replacement Adjudicator shall give written notice of acceptance of his appointment.

9.   Where an Adjudicator has already been appointed in relation to another dispute arising out of the Contract, the Chairman of

TeCSA may appoint either the same or a different person as Adjudicator.

10.   Notice requiring adjudication may be given at any time and notwithstanding that arbitration or litigation has been commenced in respect of such dispute.

11.   More than one such notice requiring adjudication may be given in respect of disputes arising out of the same contract.

AGREEMENT

12.   An agreement to adjudicate in accordance with these Rules shall be treated as an offer made by each of the Parties to TeCSA and to any Adjudicator to abide by these Rules, which offer may be accepted by conduct by appointing an Adjudicator or embarking upon the Adjudication respectively.

SCOPE OF THE ADJUDICATION

13.   The scope of the Adjudication shall be the matters identified in the notice requiring adjudication, together with

(i)   any further matters which all Parties agree should be within the scope of the Adjudication, and

(ii)   any further matters which the Adjudicator determines must be included in order that the Adjudication may be effective and/or meaningful.

14.   The Adjudicator may decide upon his own substantive jurisdiction, and as to the scope of the Adjudication.

THE PURPOSE OF THE ADJUDICATION AND THE ROLE OF THE ADJUDICATOR

15.   The underlying purpose of the Adjudication is to resolve disputes between the Parties that are within the scope of the Adjudication as rapidly and economically as is reasonably possible.

16.   Unless the Parties agree that any decisions of the Adjudicator shall be final and binding, any decision of the Adjudicator shall be binding until the dispute is finally determined by legal proceedings, by arbitration (if the Contract provides for arbitration or the parties otherwise agree to arbitration) or by agreement.

17.   Wherever possible, any decision of the Adjudicator shall reflect the legal entitlements of the Parties. Where it appears to the Adjudicator impossible to reach a concluded view upon the legal entitlements of the Parties within the practical constraints of a rapid and economical adjudication process, any decision shall represent his fair and reasonable view, in light of the facts and the law insofar as they have been

ascertained by the Adjudicator, of how the disputed matter should lie unless and until resolved by litigation or arbitration.

18.   The Adjudicator shall have the like power to open up and review any certificates or other things issued or made pursuant to the Contract as would an arbitrator appointed pursuant to the Contract and/or a court.

19.   The Adjudicator shall act fairly and impartially, but shall not be obliged or empowered to act as though he were an arbitrator.

CONDUCT OF THE ADJUDICATION

20.   The Adjudicator shall establish the procedure and timetable for the Adjudication.

21.   Without prejudice to the generality of Rule 20, the Adjudicator may if he thinks fit:

(i)   Require the delivery of written statements of case,

(ii)   Require any party to produce a bundle of key documents, whether helpful or otherwise to that Party's case, and to draw such inference as may seem proper from any imbalance in such bundle that may become apparent,

(iii)   Require the delivery to him and/or the other parties of copies of any documents other than documents that would be privileged from production to a court,

(iv)   Limit the length of any written or oral submission,

(v)   Require the attendance before him for questioning of any Party or employee or agent of any Party,

(vi)   Make site visits,

(vii)   Make use of hiw own specialist knowledge,

(viii) Obtain advice from specialist consultants, provided that at least one of the Parties so requests or consents,

(ix)   Meet and otherwise communicate with any Party without the presence of other Parties,

(x)   Make directions for the conduct of the Adjudication orally or in writing,

(xi)   Review and revise any of his own previous directions,

(xii)   Conduct the Adjudication inquisitorially, and take the initiative in ascertaining the facts and the law,

(xiii) Reach his decision(s) with or without holding an oral hearing, and with or without having endeavoured to facilitate an agreement between the Parties.

22.   The Adjudicator shall exercise such powers with a view of fairness and impartiality, giving each Party a reasonable opportunity, in light of the timetable, of putting his case and dealing with that of his opponents.

23.   The Adjudicator may not:

(i)   Require any advance payment of or security for his fees

(ii)   Receive any written submissions from one Party that are not also made available to the others,

(iii)   Refuse any Party the right at any hearing or meeting to be represented by any representative of that Party's choosing who is present,

(iv)   Act or continue to act in the face of a conflict of interest,

(v)   Subject to Rule 28, require any Party to pay or make contribution to the legal costs of another Party arising in the Adjudication

24.   The Adjudicator shall reach a decision within 28 days of referral or such longer period as is agreed by the Parties after the dispute has been referred to him. The Adjudicator shall be entitled to extend the said period of 28 days by up to 14 days with the consent of the Party by whom the dispute was referred.

ADJUDICATOR'S FEES AND EXPENSES

25.   If a Party shall request Adjudication, and it is subsequently established that he is not entitled to do so, that Party shall be solely responsible for the Adjudicator's fees and expenses.

26.   Save as aforesaid, the Parties shall be jointly responsible for the Adjudicator's fees and expenses including those of any specialist consultant appointed under Rule 19(viii). In his decision, the Adjudicator shall have the discretion to make directions with regard to those fees and expenses. If no such directions are made, the Parties shall bear such fees and expenses in equal shares, and if any Party has paid more than such equal share, that Party shall be entitled to contribution from other Parties accordingly.

27.   The Adjudicator's fees shall not exceed the rate of £1,250 per day plus expenses and VAT.

COSTS

28.   If the Parties so agree, the Adjudicator shall have jurisdiction to award costs to the successful party.

29.   Notwithstanding anything to the contrary in any contract between the Parties, the Adjudicator shall have no jurisdiction to require the Party which referred the dispute to adjudication to pay the costs of any

other Party solely by reason of having referred the dispute to adjudication.

DECISIONS

30. The Adjudicator may in any decision direct the payment of such compound or simple interest as may be commercially reasonable.

31. All decisions shall be in writing. The Adjudicator shall provide written reasons for any decision if any or all of the Parties make a request for written reasons within seven days of the date of the referral of the dispute. If requested by one Party, reasons are to be delivered to all Parties.

32.
(i) The Adjudicator may, on his own initiative or on the application of a Party, correct his decision so as to remove any clerical mistake or error arising from an accidental slip or omission;

(ii) Any application for the exercise of the Adjudicator's powers under paragraph (i) shall be made within 5 days of the date that the decision is delivered to the Parties or such shorter period as the Adjudicator may specify in his decision;

(iii) Any correction of a decision shall be made as soon as possible after the date that the application was received by the Adjudicator or, where the correction is made by the Adjudicator on his own initiative as soon as possible after he becomes aware of the need to make a correction.

ENFORCEMENT

33. Every decision of the Adjudicator shall be implemented without delay. The Parties shall be entitled to such reliefs and remedies as are set out in the decision, and shall be entitled to summary enforcement thereof, regardless of whether such decision is or is to be the subject of any challenge or review. No party shall be entitled to raise any right of set-off, counterclaim or abatement in connection with any enforcement proceedings.

IMMUNITY, CONFIDENTIALITY AND NON-COMPELLABILITY

34. Neither TeCSA, nor its Chairman, nor deputy, nor the Adjudicator nor any employee or agent of any of them shall be liable for anything done or not done in the discharge or purported discharge of his functions as Adjudicator, whether in negligence or otherwise, unless the act or omission is in bad faith.

35. The Adjudication and all matters arising in the course thereof are and will be kept confidential by the Parties except insofar as necessary to implement or enforce any decision of the Adjudicator or as may be required for the purpose of any subsequent proceedings.

36.   In the event that any Party seeks to challenge or review any decision of the Adjudicator in any subsequent litigation or arbitration, the Adjudicator shall not be joined as a party to, nor shall be summoned or otherwise required to give evidence or provide his notes in such litigation or arbitration.

LAW

37.   These Rules shall be governed by English law and under the jurisdiction of the English Courts.

38.   No Party shall, save in case of bad faith on the part of the Adjudicator, make any application to the courts whatsoever in relation to the conduct of the Adjudication or the decision of the Adjudicator until such time as the Adjudicator has made his decision, or refused to make a decision, and until the Party making the application has complied with any such decision.

October 2002

Appendix 5.2
Construction Industry Council
Model Adjudication Procedure,
Second Edition

APPENDIX 5.2
CONSTRUCTION INDUSTRY COUNCIL MODEL ADJUDICATION PROCEDURE, SECOND EDITION

General Principles

1. The object of adjudication is to reach a fair, rapid and inexpensive decision upon a dispute arising under the Contract and this procedure shall be interpreted accordingly.

   Object

2. The Adjudicator shall act impartially.

   Impartiality

3. The Adjudicator may take the initiative in ascertaining the facts and the law. He may use his own knowledge and experience. The adjudication shall be neither an arbitration nor an expert determination.

   The Adjudicator's role

4. The Adjudicator's decision shall be binding until the dispute is finally determined by legal proceedings, by arbitration (if the contract provides for arbitration or the parties otherwise agree to arbitration) or by agreement.

   Decision binding in interim

5. The Parties shall implement the Adjudicator's decision without delay whether or not the dispute is to be referred to legal proceedings or arbitration.

   Implementation of the decision

Application

6. If this procedure is incorporated into the Contract by reference, the reference shall be deemed to be to the edition current at the date of the Notice.

   Application

7. If a conflict arises between this procedure and the Contract, unless the Contract provides otherwise, this procedure shall prevail.

   Conflict

Appointment of the Adjudicator

8. Either Party may give notice at any time of its intention to refer a dispute arising under the Contract to adjudication by giving a

   Notice of adjudication

written Notice to the other Party. The Notice shall include a brief statement of the issue or issues which it is desired to refer and the redress sought. The referring Party shall send a copy of the Notice to any adjudicator named in the Contract.

Time for appointment and referral
9. The object of the procedure in paragraphs 10–14 is to secure the appointment of the adjudicator and referral of the dispute to him within 7 days of the giving of the Notice.

Appointment
10. If an adjudicator is named in the Contract, he shall within 2 days of receiving the Notice confirm his availability to act. If no adjudicator is named, or if the named adjudicator does not so confirm, the referring Party shall request the body stated in the Contract if any, or if none the Construction Industry Council, to nominate an Adjudicator within 5 days of receipt of the request. The request shall be in writing, accompanied by a copy of the Notice and the appropriate fee. Alternatively the Parties may, within 2 days of the giving of the Notice, appoint the Adjudicator by agreement.

Adjudicator unable to act
11. If, for any reason, the Adjudicator is unable to act, or fails to reach his decision within the time required by this procedure, either Party may request the body stated in the Contract if any, or if none the Construction Industry Council, to nominate a replacement adjudicator. No such request may be made after the adjudicator has notified the Parties that he has reached his decision.

Adjudicator's terms and conditions
12. Unless the Contract provides otherwise, the Adjudicator shall be appointed on the terms and conditions set out in the attached Agreement and shall be entitled to a reasonable fee and expenses.

Objection to appointment
13. If a Party objects to the appointment of a particular person as adjudicator, that objection shall not invalidate the Adjudicator's appointment or any decision he may reach.

Conduct of the Adjudication

Statement of case
14. The referring Party shall send to the Adjudicator within 7 days of the giving of the Notice (or as soon thereafter as the Adjudicator is appointed) and at the same time copy to the other Party, a statement of its case including a copy of the Notice, the Contract, details of the circumstances giving rise to the dispute, the reasons why it is entitled to the redress sought, and the evidence upon which it relies. The statement of case shall be confined to the issues raised in the Notice.

Date of referral
15. The date of referral shall be the date on which the Adjudicator receives this statement of case.

Period for decision
16. The Adjudicator shall reach his decision within 28 days of the date of referral, or such longer period as is agreed by the Parties after the

dispute has been referred. The Adjudicator may extend the period of 28 days by up to 14 days with the consent of the referring Party.

17. The Adjudicator shall have complete discretion as to how to conduct the adjudication, and shall establish the procedure and timetable, subject to any limitation there may be in the Contract or the Act. He shall not be required to observe any rule of evidence, procedure or otherwise, of any court or tribunal. Without prejudice to the generality of these powers, he may: **Procedure**

    (i)   request a written defence, further argument or counter argument

    (ii)   request the production of documents or the attendance of people whom he considers could assist

    (iii) visit the site

    (iv) meet and question the Parties and their representatives

    (v)   meet the Parties separately

    (vi) limit the length or time for submission of any statement, defence or argument

    (vii) proceed with the adjudication and reach a decision even if a Party fails to comply with a request or direction of the Adjudicator

    (viii) issue such further directions as he considers to be appropriate.

18. The Parties shall comply with any request or direction of the Adjudicator in relation to the adjudication. **Parties to comply**

19. The Adjudicator may obtain legal or technical advice, provided that he has notified the Parties of his intention first. He shall provide the Parties with copies of any written advice received. **Obtaining advice**

20 The Adjudicator shall decide the matters set out in the Notice, together with any other matters which the Parties and the Adjudicator agree shall be within the scope of the adjudication. **Matters to be determined**

21. The Adjudicator shall determine the rights and obligations of the Parties in accordance with the law of the Contract. **Adjudicator to apply the law**

22. Any Party may at any time ask that additional parties shall be joined in the adjudication. Joinder of additional parties shall be subject to the agreement of the Adjudicator and the existing and additional parties. An additional party shall have the same rights and obligations as the other Parties, unless otherwise agreed by the Adjudicator and the Parties. **Joining third parties**

23. The Adjudicator may resign at any time on giving notice in writing to the Parties. **Resignation**

The Decision

Reasons

24. The Adjudicator shall reach his decision within the time limits in paragraph 16. The Adjudicator may withhold delivery of his decision until his fees and expenses have been paid. He shall be required to give reasons unless both Parties agree at any time that he shall not be required to give reasons.

Late decisions

25. If the Adjudicator fails to reach his decision within the time permitted by this procedure, his decision shall nonetheless be effective if reached before the referral of the dispute to any replacement adjudicator under paragraph 11 but not otherwise. If he fails to reach such an effective decision, he shall not be entitled to any fees or expenses (save for the cost of any legal or technical advice subject to the Parties having received such advice).

Power to open up certificates

26. The Adjudicator may open up, review and revise any certificate, decision, direction, instruction, notice, opinion, requirement or valuation made in relation to the Contract.

Interest

27. The Adjudicator may in any decision direct the payment of such simple or compound interest from such dates, at such rates and with such rests, as he considers appropriate.

Costs

28. The Parties shall bear their own costs and expenses incurred in the adjudication.

Adjudicator's fees and expenses

29. The Parties shall be jointly and severally liable for the Adjudicator's fees and expenses, including those of any legal or technical adviser appointed under paragraph 19, but the Adjudicator may direct a Party to pay all or part of the fees and expenses. If he makes no such direction, the Parties shall pay them in equal shares. The Party requesting the adjudication shall be liable for the Adjudicator's fees and expenses if the adjudication does not proceed.

Enforcement

30. The Parties shall be entitled to the redress set out in the decision and to seek summary enforcement, whether or not the dispute is to be finally determined by legal proceedings or arbitration. No issue decided by the Adjudicator may subsequently be referred for decision by another adjudicator unless so agreed by the Parties.

Subsequent decision by arbitration or court

31. In the event that the dispute is referred to legal proceedings or arbitration, the Adjudicator's decision shall not inhibit the right of the court or arbitrator to determine the Parties' rights or obligations as if no adjudication had taken place.

Miscellaneous Provisions

Adjudicator not to be appointed arbitrator

32. Unless the Parties agree, the Adjudicator shall not be appointed arbitrator in any subsequent arbitration between the Parties under the Contract. No Party may call the Adjudicator as a witness in any

legal proceedings or arbitration concerning the subject matter of the adjudication.

33. The Adjudicator is not liable for anything done or omitted in the discharge or purported discharge of his functions as Adjudicator (whether in negligence or otherwise) unless the act or omission is in bad faith, and any employee or agent of the Adjudicator is similarly protected from liability. **Immunity of the Adjudicator**

34. The Adjudicator is appointed to determine the dispute or disputes between the Parties and his decision may not be relied upon by third parties, to whom he shall owe no duty of care. **Reliance**

35. This procedure shall be interpreted in accordance with the law of England and Wales. **Proper law**

Definitions

'Act' means the Housing Grants, Construction and Regeneration Act 1996.

'Adjudicator' means the person named as such in the Contract or appointed in accordance with this Procedure.

'Contract' means the contract between the Parties which contains the provision for adjudication.

'Notice' means the notice given under paragraph 3.

'Party' means a party to the Contract, and any additional parties joined under paragraph 22. 'Referring Party' means the Party who gives notice under paragraph 8.

AGREEMENT

THIS AGREEMENT is made on the . . . . day of . . . . 20 . . . . . .

Between

1. . . . .

   of . . . .

   . . . . (the referring Party)

2. . . . .

   of . . . .

   . . . . (the other Party)

(together called the Parties)

3. . . . .

   of . . . .

   . . . . (the Adjudicator)

A dispute has arisen between the Parties under a Contract between them dated . . . .

. . . .

in connection with . . . .

This dispute has been referred to adjudication in accordance with the CIC Model Adjudication Procedure (the Procedure) and the Adjudicator has been requested to act.

IT IS AGREED that

1. The rights and obligations of the Adjudicator and the Parties shall be as set out in this Agreement.

2. The Adjudicator agrees to adjudicate the dispute in accordance with the Procedure.

3. The Parties agree jointly and severally to pay the Adjudicator's fees and expenses as set out in the attached schedule and in accordance with the Procedure.

4. The Adjudicator and the Parties shall keep the adjudication confidential, except so far as is necessary to enable a Party to implement or enforce the Adjudicator's decision.

5. The Parties acknowledge that the Adjudicator shall not be liable for anything done or omitted in the discharge or purported discharge of his functions as Adjudicator (whether in negligence or otherwise) unless the act or omission is in bad faith, and any employee or agent of the Adjudicator shall be similarly protected from liability.

6. This Agreement shall be interpreted in accordance with the law of England and Wales.

Schedule

1. The Adjudicator shall be paid £........ per hour in respect of all time spent on the adjudication, including travelling time with a maximum of £............. per day.

2. The Adjudicator shall be reimbursed his reasonable expenses and disbursements, in respect of travelling, hotel and similar expenses, room charges, the cost of legal or technical advice obtained in accordance with the Procedure and other extraordinary expenses necessarily incurred.

3. The Adjudicator is/is not * currently registered for VAT. Where the Adjudicator is registered for VAT, it shall be charged additionally in accordance with the rates current at the datte of the work done.

* delete as applicable.

Signed on behalf of the referring Party

. . . .

Signed on behalf of the other Party

. . . .

Signed on behalf of the Adjudicator

. . . .

Appendix 6
Adjudicator Nominating Bodies

APPENDIX 6
ADJUDICATOR NOMINATING BODIES

Academy of Independent Construction Adjudicators
Centre for Dispute Resolution
Chartered Institute of Arbitrators
Chartered Institute of Arbitrators (Scotland)
Chartered Institute of Building
Construction Confederation
Construction Employers Confederation (Northern Ireland)
Construction Industry Council
Institution of Chemical Engineers
Institution of Electrical Engineers
Institution of Civil Engineers
Institution of Mechanical Engineers
Royal Institute of British Architects
Royal Institution of Chartered Surveyors
Royal Institution of Chartered Surveyors (Scotland)
Technology and Construction Solicitors Association

Appendix 7
Alphabetical List of Cases

APPENDIX 7
ALPHABETICAL LIST OF CASES

Reported cases and known unreported cases as at 31 March 2003

S denotes Scottish Case
CoA denotes Court of Appeal
AC denotes Appeal Court (Scotland)

1. A & D Maintenance & Construction Ltd *v.* Pagehurst Construction Services Ltd (June 1999) 30, 70
2. ABB Zantingh Ltd *v.* Zedal Building Services Ltd (Dec 2000) 26
3. ABB Power Construction Ltd *v.* Norwest Holst Engineering Ltd (Aug 2000) 27
4. Absolute Rentals Ltd *v.* Gencor Enterprises Ltd (Jan 2000) ... 39, 70, 93
5. Allied London & Scottish Properties plc *v.* Riverbrae Construction Ltd (July 1999) **S** 71
6. Atlas (The) Ceiling & Partition Co Ltd *v.* Crowngate Estates (Cheltenham) Ltd (Feb 2000) 22
7. Austin Hall Building Ltd *v.* Buckland Securities Ltd (April 2001) ... 114
8. Balfour Kilpatrick Ltd *v.* Glauser International SA (July 2000) 31
9. Balfour Beatty Construction Ltd *v.* The Mayor & Burgesses of the London Borough of Lambeth (April 2002) 100, 106
10. Ballast Plc *v.* The Burrell Company (Construction Management) Ltd (June 2001) (1) **S** 65, 108
11. Barr Ltd *v.* Law Mining Ltd (June 2001) **S** 31, 115
12. Bickerton Construction Ltd *v.* Temple Windows Ltd (June 2001) ... 23
13. Birchall (Patrick PA) t/a Pier 1 Metalworks *v.* West Morland Car Sales Ltd (Nov 2001) 114
14. Bloor Construction (UK) Ltd *v.* Bowmer & Kirkland (London) Ltd (April 2000) 111
15. Bouygues UK Ltd *v.* Dahl-Jensen UK Ltd (Nov 1999) (1) 109, 111
16. Bouygues UK Ltd *v.* Dahl-Jensen UK Ltd (CoA July 2000) (2) ... 71, 93
17. Bovis Lend Lease Ltd *v.* Triangle Development Ltd (Nov 2002) 70
18. Britcon (Scunthorpe) Ltd *v.* Lincolnsfields Ltd (Aug 2001) 65
19. Bridgeway Construction Ltd *v.* Tolent Construction Ltd (April 2000) 40

20. British Waterways Board (Petition of) (July 2001) **S** 20
21. Butler (Tim) Contractors Ltd *v*. Merewood Homes Ltd (April 2000)
. . . . 19, 122
22. C & B Scene Concept Design Ltd *v*. Isobars Ltd (June 2001) (1) 111
23. C & B Scene Concept Design Ltd *v*. Isobars Ltd (CoA Jan 2002) (2)
. . . . 111
24. Canary Riverside Development *v*. Timtec International (Nov 2000)
. . . . 33
25. Carter (RG) *v*. Edmund Nuttall Ltd (1) (June 2000) 33
26. Carter (RG) *v*. Edmund Nuttall Ltd (2) (April 2002) 82, 99
27. Chamberlain Carpentry & Joinery *v*. Alfred McAlpine Construction
Ltd (March 2002) 31, 39
28. Christiani & Nielsen Ltd *v*. The Lowry Centre & Development Co
Ltd (June 2000) 22, 32, 78
29. City Inn Ltd *v*. Shepherd Construction Ltd (July 2001) **S** 117
30. Clark Contracts Ltd *v*. The Burrell Company (Construction
Management) Ltd (Jan 2002) **S** 125
31. Construction Centre Group Ltd *v*. The Highland Council (Aug 2002)
. . . . 70
32. Cook (FW) Ltd *v*. Shimizu (UK) Ltd (Feb 2000) 62
33. Cothliff (John) Ltd *v*. Allen Build (North West) Ltd (July 1999) 39
34. Discain Project Services Ltd *v*. Opecprime Development Ltd (1)
(Aug 2000) 20, 100
35. Discain Project Services Ltd *v*. Opecprime Development Ltd (2)
(Apr 2001) 100, 105, 114
36. Durabella Ltd *v*. J. Jarvis & Sons Ltd (Sept 2001) 127
37. Earls Terrace Properties Ltd *v*. Waterloo Investments Ltd (Feb 2002)
. . . . 22
38. Elanay Contracts Ltd *v*. The Vestry (Aug 2000) 114
39. Edinburgh Royal Joint Venture *v*. Broderick Structures Ltd
(Aug 2002) 155
40. Faithful & Gould Ltd *v*. Arcal Ltd and Others (May 2001) 111
41. Farebrother Building Services Ltd *v*. Frogmore Investments Ltd
(April 2001) 65, 78
42. Fastrack Contractors Ltd *v*. Morrison Construction Ltd, Impreglio
UK Ltd (Jan 2000) 20, 21, 32, 42
43. Fence Gate Ltd *v*. James R. Knowles Ltd (June 2001) 24, 83
44. Finney (Joseph) plc *v*. Gordon Vickers & Gary Vickers t/a Mill Hotel
(March 2001) 32
45. Gibson Lea Retail Interiors Ltd *v*. Makro Self Service Wholesalers Ltd
(July 2001) 25
46. Gibson *v*. Imperial Homes (Feb 2002) 19
47. Gillies Ramsey Diamond *v*. PJW Enterprises Ltd (June 2002) **S** 24

48. Glencot Development & Design Co Ltd *v.* Ben Barrett & Son (Contractors) Ltd (Feb 2001) 114, 153

49. Griffin (Ken) & John Tomlinson t/a K & D Contractors *v.* Midas Homes Ltd (July 2000) 42

50. Grovedeck Ltd *v.* Capital Demolition Ltd (Feb 2000) 23, 30, 115

51. Guardi Shoes *v.* Datum Contracts International Ltd (Oct 2002) 124

52. Harwood Construction Ltd *v.* Lantrode Ltd (Nov 2000) 30

53. Herschel Engineering Ltd *v.* Breen Property Ltd (1) (April 2000) ... 30

54. Herschel Engineering Ltd *v.* Breen Property Ltd (2) (July 2000) 65, 69, 93

55. Holt Insulation Ltd *v.* Colt International Ltd (Feb 2001) 32

56. Homer Burgess Ltd *v.* Chirex (Annan) Ltd (Nov 1999) S 27, 78

57. Impresa Castelli SPA *v.* Cola Holdings Ltd (May 2002) 117

58. Jenson (Paul) Ltd *v.* Staveley Industries plc (Sept 2001) 38, 113

59. Jerome Engineering Ltd *v.* Lloyd Morris Electrical Ltd (Nov 2001) 42, 84

60. Karl Construction (Scotland) Ltd *v.* Sweeney Civil Engineering (Scotland) Ltd (Dec 2000) (1) S 85, 100, 121

61. Karl Construction (Scotland) Ltd *v.* Sweeney Civil Engineering (Scotland) Ltd (AC Jan 2002) (2) S 121

62. KNS Industrial Services (Birmingham) Ltd *v.* Sindall Ltd (July 2000) 42, 85

63. Lathom Construction Ltd *v.* Brian Cross and Anne Cross (Oct 1999) 32

64. Levolux A.T. Ltd *v.* Ferson Contractors Ltd (June 2002) (-) 70

65. LPL Electrical Services Ltd *v.* Kershaw Mechanical Services Ltd (Feb 2001) 54

66. Macob Civil Engineering Ltd *v.* Morrison Construction Ltd (Feb 1999) 67, 100

67. Mackley (JT) & Co Ltd *v.* Gosport Marina Ltd (July 2002) 117

68. Maxi Construction Management Ltd *v.* Morton Rolls Ltd (Aug 2001) 123

69. Maymac Environmental Services Ltd *v.* Faraday Building Services Ltd (July 2000) 21, 79

70. McAlpine (Sir Robert) *v.* Pring & St Hill Ltd (Oct 2001) 70

71. McLean (David) Housing Contractors Ltd *v.* Swansea Housing Association Ltd (July 2002) 31, 70

72. Mecright Ltd *v.* TA Morris Developments Ltd (June 2001) 41

73. Millers Specialist Joinery Company Ltd *v.* Nobles Construction Ltd (Aug 2001) 125

74. Mitsui Babcock Energy Services Ltd (Petition of) (June 2001) S ... 27

75. Mowlem (John) & Co plc *v.* Hydra-Tight & Co plc t/a Hevilifts (June 2000) 33, 137

76. Naylor (William) *v.* Greenacres Curling Ltd (June 2001) **S** 31
77. Nolan Davis Ltd *v.* Steven P. Catton (Feb 2000) 19, 39, 79
78. Nordot Engineering Services Ltd *v.* Siemens plc (April 2000) 78
79. Northern Developments (Cumbria) Ltd *v.* J & J Nichol (Jan 2000)
      .... 18, 30, 39, 70, 125
80. Nottingham Community Housing Association Ltd *v.* Powerminster
      Ltd (June 2000) 28
81. Nuttall (Edmund) Ltd *v.* Sevenoaks District Council (April 2000)
      .... 70, 111, 124
82. Nuttall (Edmund) *v.* RG Carter Ltd (March 2002) 33, 51, 85, 104
83. Oakley (William) and David Oakley *v.* Airclear Environmental Ltd
      and Airclear TS Ltd (Oct 2001) 19
84. Outwing Construction Ltd *v.* H. Randall and Son Ltd (March 1999)
      .... 67, 68
85. Palmers Ltd *v.* ABB Power Construction Ltd (Aug 1999) 27, 124
86. Parsons Plastics (Research & Development) Ltd *v.* Purac Ltd
      (Aug 2001) (1) 21
87. Parsons Plastics (Research & Development) Ltd *v.* Purac Ltd
      (CoA April 2002) (2) 70
88. Parke (George) *v.* The Fenton Gretton Partnership (Aug 2000)
      .... 69, 71, 93, 144
89. Pro Design Ltd *v.* New Millenium Experience Company Ltd
      (Sept 2001) 69
90. Project Consultancy Group *v.* Trustees of the Gray Trust (July 1999)
      .... 21, 78
91. Quality Street Properties (Trading) Ltd *v.* Elmwood (Glasgow) Ltd
      (Feb 2002) **S** 30, 31
92. Rainford House Ltd *v.* Cadogan Ltd (Feb 2001) 71, 93
93. Re A Company (GAL *v.* CCL) (May 2001) 1299 of 2001 68
94. Rentokil Ailsa Environmental Ltd *v.* Eastend Civil Engineering Ltd
      (March 1999) **S** 30
95. RJT Consulting Engineers Ltd *v.* DM Engineering (Northern Ireland)
      Ltd (May 2001) (1) 23
96. RJT Consulting Engineers Ltd *v.* DM Engineering (Northern Ireland)
      Ltd (CoA Mar 2002) (2) 24
97. Samuel Thomas Construction Ltd *v.* Bick & Bick (aka J & B
      Development) (Jan 2000) 28
98. Shepherd Construction Ltd *v.* Mecright Ltd (July 2000) 32
99. Sherwood & Casson Ltd *v.* McKenzie (Nov 1999) 31
100. Shimizu Europe Ltd *v.* Automajor Ltd (2002) 20, 111
101. Sindall Ltd *v.* Solland (June 2001) 20
102. Skansa Construction UK Ltd *v.* ERDC Group and John Hunter
      (Nov 2002) **S** 32

103. SL Timber Systems Ltd *v.* Carillion Construction Ltd (June 2001) **S**
 . . . . 93, 125
104. Solland International Ltd *v.* Daraydan Holdings Ltd (Feb 2002) 71
105. Staveley Industries plc t/a EI.WHS *v.* Odebrecht Oil & Gas Services
 Ltd (Feb 2001) 25, 27
106. Stiell Ltd *v.* Reima Control Systems Ltd (June 2000) **S** 30
107. Strathmore Building Services Ltd *v.* Colin Scott Grieg t/a Hestia
 Fireside Design (May 2000) **S** 124
108. Straume (A) UK Ltd *v.* Bradlor Developments Ltd (April 1999)
 . . . . 33, 69
109. Stubbs Rich Architects *v.* W.H. Tolley & Son Ltd (Aug 2001) . . . 38, 112
110. Total M & E Services Ltd *v.* ABB Building Technologies Ltd
 (Feb 2002) 39
111. Trentham (Barry D) *v.* Lawfield Investments Ltd (May 2002) **S** . . . 68
112. Universal Music Operations Ltd *v.* Flairnote Ltd (Aug 2000) 40
113. VHE Construction plc *v.* RBSTB Trust Company Ltd (Jan 2000)
 . . . . 31, 70, 108, 124, 125
114. Watkin Jones & Son Ltd *v.* Lidl UK GMBH (Dec 2001) (1) 79
115. Watkin Jones & Son Ltd *v.* Lidl UK GMBH (Feb 2002) (2) 32
116. Watson Building Services Ltd *v.* Miller (Preservation) Ltd
 (Mar 2001) **S** 19, 32
117. Whiteways Contractors (Sussex) Ltd *v.* Impresa Castelli Construction
 UK Ltd (Aug 2000) 71, 78, 124, 125
118. Woods Hardwick Ltd *v.* Chiltern Air-Conditioning Ltd (July 2000)
 . . . . 100, 125
119. Workplace Technologies plc *v.* E. Squared Ltd and Mr J.L. Riches
 (Feb 2000) 22
120. Yarm Road Ltd *v.* Costain Ltd (July 2001) 22

Recent additions at time of going to press

121. A *v.* B (Dec 2002) **S** 68
122. Ashley House plc *v.* Galliers Southern Ltd (Feb 2002) 93
123. Baldwin Industrial Services *v.* Barr Ltd (Dec 2002) 26, 71
124. Ballast Plc *v.* The Burrell Company (Construction Management) Ltd
 (AC Dec 2002) (2) **S** 65, 108
125. Brenton *v.* Palmer (Jan 2001) 19
126. Carillion Construction Ltd *v.* Devonport Royal Dockyard Ltd
 (Nov 2002) 20, 23
127. Cowlin Contracts Ltd *v.* CFW Architects (Nov 2002) 19, 63
128. Debeck Ductwork Insulation Ltd *v.* T & E Engineers Ltd (Oct 2002)
 24

Ferson Construction Ltd *v.* Levolux AT Ltd (See[132] Levolux AT Ltd *v.* Ferson Construction Ltd)

129. Gennaro Maurizi Picardi t/a Picardi Architects *v.* Cuniberti and Another (Dec 2002) 21, 28

130. Hart Builders (Edinburgh) Ltd *v.* St Andrews Ltd (Aug 2002) **S** 124

131. Joinery Plus Ltd (in administration) *v.* Laing Ltd (CoA Jan 2003) 111

132. Levolux AT Ltd *v.* Ferson Construction Ltd (CoA Jan 2003) (2) 70, 155

133. Mivan Ltd *v.* Lighting Technology Projects Ltd (April 2001) 32

134. Try Construction Ltd *v.* Eton Town House Group Ltd (Jan 2003) 107, 125

Cases included in Categorised List (Appendix 8) but not referred to in text

R. Durtnell & Sons Ltd *v.* Kaduna Ltd (March 2003)

Hitec Power Protection BV *v.* MCI Worldcom Ltd (Aug 2002)

Martin Girt *v.* Page Bentley (2002)

Pegram *v.* Tally Weijl (2003)

Trustees of the Harbour of Peterhead *v.* Lilley Construction Ltd (April 2003) **S**

Appendix 8
Categorised List of Cases

APPENDIX 8
CATEGORISED LIST OF CASES

Most judgments lay down principles of general application but care should be taken before reliance is placed on a particular judgment as some arise out of the particular circumstances of the case and consequently may not be wholly of general application. This does not always emerge in the short commentaries which appear in the press and on the internet. A comparison of case summaries available from various sources will also demonstrate that different commentators do not always emphasise the same aspect of a judgment. If a case appears relevant to a situation the full judgment should be studied before reliance is placed upon it.

Cases are categorised under the aspect or aspects for which they are generally best known. A case may therefore appear in more than one category. At the same time a judgment may be of relevance in respect of additional aspects for which it is not listed. Cases listed are reported cases and known unreported cases as at 31 March 2003. Most of the cases are referred to in the text, as indicated in Appendix 7.

Obstructions to the right to immediate adjudication

A *v.* B
Carter (RG) *v.* Edmund Nuttall Ltd (1)
Christiani & Neilson Ltd *v.* The Lowry Centre & Development Co Ltd
Impresa Castelli SPA *v.* Cola Holdings Ltd
Mowlem (John) & Co plc *v.* Hydra-Tight & Co plc t/a Hevilifts
Pegram *v.* Tally Weijl
Trustees of the Harbour of Peterhead *v.* Lilley Construction Ltd

The contract

Evidence in writing

Birchall (Patrick PA) t/a Pier 1 Metalworks *v.* West Morland Car Sales Ltd
Carillion Construction Ltd *v.* Devonport Royal Dockyard Ltd
Debeck Ductwork Insulation Ltd *v.* T & E Engineers Ltd (Oct 2002)
Grovedeck Ltd *v.* Capital Demolition Ltd

Millers Specialist Joinery Company *v.* Nobles Construction Ltd
RJT Consulting Engineers Ltd *v.* DM Engineering (NI) Ltd (1) (2)
Strathmore Building Services Ltd *v.* Colin Scott Grieg t/a Hestia Fireside Design

Evidence of a contract/contract terms

Brenton *v.* Palmer
Butler (Tim) Contractors Ltd *v.* Merewood Homes Ltd
Cowlin Contracts Ltd *v.* CFW Architects
Gibson *v.* Imperial Homes
Oakley (William) and David Oakley *v.* Airclear Environmental Ltd and Airclear TS Ltd
Nolan Davis Ltd *v.* Steven P. Catton
Watson Building Services Ltd *v.* Miller (Preservation) Ltd

Termination of contract

A & D Maintenance & Construction Ltd *v.* Pagehurst Construction Services Ltd
Impresa Castelli SPA *v.* Cola Holdings Ltd
KNS Industrial Services (Birmingham) Ltd *v.* Sindall Ltd
Northern Developments (Cumbria) Ltd *v.* J & J Nichol

Date contract concluded

Atlas (The) Ceiling & Partition Company Ltd *v.* Crowngate Estates (Cheltenham) Ltd
Christiani & Neilson Ltd *v.* The Lowry Centre & Development Co Ltd
Earls Terrace Properties Ltd *v.* Waterloo Investments Ltd
Project Consultancy Group *v.* Trustees of the Gray Trust
Workplace Technologies plc *v.* E Squared Ltd and Mr J. L. Riches
Yarm Road Ltd *v.* Costain Ltd

Interpretation of sections 104–107 of the Act

ABB Power Construction Ltd *v.* Norwest Holst Engineering Ltd
ABB Zantingh Ltd *v.* Zedal Building Services Ltd
Baldwin Industrial Services *v.* Barr Ltd
Earls Terrace Properties Ltd *v.* Waterloo Investments Ltd
Fence Gate Ltd *v.* James R. Knowles Ltd

Gibson Lea Retail Interiors Ltd *v.* Makro Self Service Wholesalers Ltd
Gillies Ramsey Diamond *v.* PJW Enterprises Ltd
Homer Burgess Ltd *v.* Chirex (Annan) Ltd
Mitsui Babcock Energy Services Ltd (Petition of)
Nordot Engineering Services Ltd *v.* Siemens plc
Nottingham Community Housing Association Ltd *v.* Powerminster Ltd
Palmers Ltd *v.* ABB Power Construction Ltd
Samuel Thomas Construction Ltd *v.* Bick & Bick (aka J & B Development)
Staveley Industries plc t/a EI.WHS *v.* Odebrecht Oil & Gas Services Ltd

Mixed contracts

Fence Gate Ltd *v.* James R. Knowles Ltd
Homer Burgess Ltd *v.* Chirex (Annan) Ltd

The dispute

Nature and evidence of a dispute

British Waterways Board (Petition of)
Discain Project Services Ltd *v.* Opecprime Development Ltd (2)
R. Durtnell & Sons Ltd *v.* Kaduna Ltd
Fastrack Contractors Ltd *v.* Morrison Construction Ltd, Impreglio UK Ltd
Gillies Ramsey Diamond *v.* PJW Enterprises Ltd
Hitec Power Protection BV *v.* MCI Worldcom Ltd
KNS Industrial Services (Birmingham) Ltd *v.* Sindall Ltd
Nuttall (Edmund) *v.* RG Carter Ltd
Sindall Ltd *v.* Solland

Duplication

Holt Insulation Ltd *v.* Colt International Ltd
Mivan Ltd *v.* Lighting Technology Projects Ltd
Naylor (William) *v.* Greenacres Curling Ltd
Sherwood & Casson Ltd *v.* McKenzie
Skansa Construction UK Ltd *v.* ERDC Group and John Hunter
VHE Construction Plc *v.* RBSTB Trust Company Ltd
Watkin Jones & Son Ltd *v.* Lidl UK GMBH (2)

Multiple disputes

Balfour Kilpatrick Ltd *v.* Glauser International SA
Barr Ltd *v.* Law Mining Ltd
Chamberlain Carpentry & Joinery Ltd *v.* Alfred McAlpine Construction Ltd
Fastrack Contractors Ltd *v.* Morrison Construction Ltd, Impreglio UK Ltd
Grovedeck Ltd *v.* Capital Demolition Ltd
McLean (David) Housing Contractors Ltd *v.* Swansea Housing Association Ltd
Quality Street Properties (Trading) Ltd *v.* Elmwood (Glasgow) Ltd

Notice and referral

Bickerton Construction Ltd *v.* Temple Windows Ltd
Carter (RG) *v.* Edmund Nuttall Ltd (2)
Cook (FW) Ltd *v.* Shimizu (UK) Ltd
Fastrack Contractors Ltd *v.* Morrison Construction Ltd, Impreglio UK Ltd
Griffin (Ken) & John Tomlinson t/a K & D Contractors *v.* Midas Homes Ltd
Jerome Engineering Ltd *v.* Lloyd Morris Electrical Ltd
KNS Industrial Services (Birmingham) Ltd *v.* Sindall Ltd
LPL Electrical Services Ltd *v.* Kershaw Mechanical Services Ltd
Northern Developments (Cumbria) Ltd *v.* J & J Nichol
Strathmore Building Services Ltd *v.* Colin Scott Grieg t/a Hestia Fireside Design
Whiteways Contractors (Sussex) Ltd *v.* Impresa Castelli Construction UK Ltd

Settlements

Finney (Joseph) plc *v.* Gordon Vickers & Gary Vickers t/a Mill Hotel
Lathom Construction Ltd *v.* Brian Cross and Anne Cross
Quality Street Properties (Trading) Ltd *v.* Elmwood (Glasgow) Ltd
Shepherd Construction Ltd *v.* Mecright Ltd

Concurrent court or arbitration proceedings

Absolute Rentals Ltd *v.* Gencor Enterprises Ltd
Baldwin Industrial Services *v.* Barr Ltd
Harwood Construction Ltd *v.* Lantrode Ltd
Herschel Engineering Ltd *v.* Breen Property Ltd (1)
Quality Street Properties (Trading) Ltd *v.* Elmwood (Glasgow) Ltd
Rentokil Ailsa Environmental Ltd *v.* Eastend Civil Engineering Ltd
Stiell Ltd *v.* Reima Control Systems Ltd

Jurisdiction

Adjudicator deciding his own

Brenton *v*. Palmer
Christiani & Neilson Ltd *v*. The Lowry Centre & Development Co Ltd
Cowlin Contracts Ltd *v*. CFW Architects (Nov 2002)
Farebrother Building Services Ltd *v*. Frogmore Investments Ltd
Homer Burgess Ltd *v*. Chirex (Annan) Ltd
Maymac Environmental Services Ltd *v*. Faraday Building Services Ltd
Nolan Davis Ltd *v*. Steven P. Catton
Nordot Engineering Services Ltd *v*. Siemens plc
Project Consultancy Group *v*. Trustees of the Gray Trust
Watson Building Services Ltd *v*. Miller (Preservation) Ltd
Whiteways Contractors (Sussex) Ltd *v*. Impresa Castelli Construction UK Ltd

Excess of jurisdiction

Barr Ltd *v*. Law Mining Ltd
Bickerton Construction Ltd *v*. Temple Windows Ltd
C & B Scene Concept Design Ltd *v*. Isobars Ltd (2)
Griffin (Ken) & John Tomlinson t/a K & D Contractors *v*. Midas Homes Ltd
LPL Electrical Services Ltd *v*. Kershaw Mechanical Services Ltd
Mecright Ltd *v*. TA Morris Developments Ltd

Failure to challenge

Glencot Development & Design Co Ltd *v*. Ben Barrett & Son (Contractors) Ltd
Maymac Environmental Services Ltd *v*. Faraday Building Services Ltd

Objection to adjudicator

Carter (RG) *v*. Edmund Nuttall Ltd (2)

Contra-claims

Set-off

Bovis Lend Lease Ltd *v*. Triangle Development Ltd
Harwood Construction Ltd *v*. Lantrode Ltd

Levolux A.T. Ltd *v.* Ferson Contractors Ltd (1) (2)
McAlpine (Sir Robert) *v.* Pring & St Hill Ltd
McLean (David) Housing Contractors Ltd *v.* Swansea Housing Association
  Ltd
Millers Specialist Joinery Company Ltd *v.* Nobles Construction
Northern Developments (Cumbria) Ltd *v.* J & J Nichol
Nuttall (Edmund) Ltd *v.* Sevenoaks District Council
Parsons Plastics (Research & Development) Ltd *v.* Purac Ltd (1) (2)
VHE Construction plc *v.* RBSTB Trust Company Ltd

Counterclaim

A & D Maintenance & Construction Ltd *v.* Pagehurst Construction Services
  Ltd
Absolute Rentals Ltd *v.* Gencor Enteprises Ltd
Allied London & Scottish Properties plc *v.* Riverbrae Construction Ltd
Construction Centre Group Ltd *v.* The Highland Council
Farebrother Building Services Ltd *v.* Frogmore Investments Ltd
Rainford House Ltd *v.* Cadogan Ltd
Solland International Ltd *v.* Daraydan Holdings Ltd
Total M & E Services Ltd *v.* ABB Building Technologies Ltd
Whiteways Contractors (Sussex) Ltd *v.* Impresa Castelli Construction UK Ltd

Financial circumstances of the parties

Ashley House plc *v.* Galliers Southern Ltd
Absolute Rentals Ltd *v.* Gencor Enterprises Ltd
Baldwin Industrial Services *v.* Barr Ltd
Bouygues UK Ltd *v.* Dahl-Jensen UK Ltd (2)
Canary Riverside Development *v.* Timtec International
Harwood Construction Ltd *v.* Lantrode Ltd
Herschel Engineering Ltd *v.* Breen Property Ltd (2)
Karl Construction (Scotland) Ltd *v.* Sweeney Civil Engineering (Scotland) Ltd
  (2)
Parke (George) *v.* The Fenton Gretton Partnership
Rainford House Ltd *v.* Cadogan Ltd
Re A Company GAL *v.* CCL 1299 of 2001
SL Timber Systems Ltd *v.* Carillion Construction Ltd
Trentham (Barry D) *v.* Lawfield Investments Ltd
Straume (A) UK Ltd *v.* Bradlor Developments Ltd

Adjudicator's conduct of the adjudication

Procedural fairness, natural justice

Balfour Beatty Construction Ltd *v.* The Mayor & Burgesses of the London Borough of Lambeth

Discain Project Services Ltd *v.* Opecprime Development Ltd (1) (2)

Glencot Development & Design Co Ltd *v.* Ben Barrett & Son (Contractors) Ltd

Karl Construction (Scotland) Ltd *v.* Sweeney Civil Engineering (Scotland) Ltd (1)

Mecright Ltd *v.* TA Morris Developments Ltd

Try Construction Ltd *v.* Eton Town House Group Ltd (Jan 2003)

Woods Hardwick Ltd *v.* Chiltern Air Conditioning Ltd

Human Rights Act

Austin Hall Building Ltd *v.* Buckland Securities Ltd

Birchall (Patrick PA) t/a Pier 1 Metalworks *v.* West Morland Car Sales Ltd

Elanay Contracts Ltd *v.* The Vestry

The decision

Principles of enforcement

Balfour Beatty Construction Ltd *v.* The Mayor & Burgesses of the London Borough of Lambeth

C & B Scene Concept Design Ltd *v.* Isobars Ltd (2)

City Inn Ltd *v.* Shepherd Construction Ltd

R. Durtnell & Sons Ltd *v.* Kaduna Ltd

Edinburgh Royal Joint Venture *v.* Broderick Structures Ltd

Macob Civil Engineering Ltd *v.* Morrison Construction Ltd

Outwing Construction Ltd *v.* H. Randall & Son Ltd

Pro Design Ltd *v.* New Millenium Experience Company Ltd

Rentokil Ailsa Environmental Ltd *v.* Eastend Civil Engineering Ltd

Shimizu Europe Ltd *v.* Automajor Ltd

Correction of slips

Bloor Construction (UK) Ltd *v.* Bowmer & Kirkland (London) Ltd

Edinburgh Royal Joint Venture *v.* Broderick Structures Ltd
Nuttall (Edmund) Ltd *v.* Sevenoaks District Council

Mistakes and errors

Baldwin Industrial Services *v.* Barr Ltd
Ballast Plc *v.* The Burrell Company (Construction Management) Ltd (1) (2)
Bloor Construction (UK) Ltd *v.* Bowmer & Kirkland (London) Ltd
Bouygues UK Ltd *v.* Dahl-Jensen UK Ltd (1)
Britcon (Scunthorpe) Ltd *v.* Lincolnsfields Ltd (Aug 2001)
C & B Scene Concept Design Ltd *v.* Isobars Ltd (1) (2)
Farebrother Building Services Ltd *v.* Frogmore Investments Ltd
Joinery Plus Ltd (in administration) *v.* Laing Ltd
Nuttall (Edmund) Ltd *v.* Sevenoaks District Council
Shimizu Europe Ltd *v.* Automajor Ltd
SL Timber Systems Ltd *v.* Carillion Construction Ltd

Costs

Adjudicator's costs

Faithful & Gould Ltd *v.* Arcal Ltd and Others
Jenson (Paul) Ltd *v.* Staveley Industries plc
Stubbs Rich Architects *v.* W.H. Tolley & Son Ltd

Parties' costs

Absolute Rentals Ltd *v.* Gencor Enterprises Ltd
Bridgeway Construction Ltd *v.* Tolent Construction Ltd
Chamberlain Carpentry & Joinery Ltd *v.* Alfred McAlpine Construction Ltd
Cothliff (John) Ltd *v.* Allen Build (North West) Ltd
Nolan Davis Ltd *v.* Steven P. Catton
Northern Developments (Cumbria) Ltd *v.* J & J Nichol
Total M & E Services Ltd *v.* ABB Building Technologies Ltd

Payment matters

Butler (Tim) Contractors Ltd *v.* Merewood Homes Ltd
Clark Contracts Ltd *v.* The Burrell Company (Construction Management) Ltd

Durabella Ltd *v*. J. Jarvis & Sons Ltd
Edinburgh Royal Joint Venture *v*. Broderick Structures Ltd
Guardi Shoes *v*. Datum Contracts International Ltd
Hart Builders (Edinburgh) Ltd *v*. St Andrews Ltd
Karl Construction (Scotland) Ltd *v*. Sweeney Civil Engineering (Scotland) Ltd (1)
LPL Electrical Services Ltd *v*. Kershaw Mechanical Services Ltd
Martin Girt *v*. Page Bentley
Maxi Construction Management Ltd *v*. Morton Rolls Ltd
Millers Specialist Joinery Company Ltd *v*. Nobles Construction Ltd
Mowlem (John) & Co. plc *v*. Hydra-Tight & Co plc t/a Hevilifts
Northern Developments (Cumbria) Ltd *v*. J & J Nichol
Palmers Ltd *v*. ABB Power Construction Ltd
SL Timber Systems Ltd *v*. Carillion Construction Ltd
Strathmore Building Services Ltd *v*. Colin Scott Grieg t/a Hestia Fireside Design
VHE Construction plc *v*. RBSTB Trust Company Ltd
Watkin Jones & Son Ltd *v*. Lidl UK GMBH (1)
Whiteways Contractors (Sussex) Ltd *v*. Impresa Castelli Construction UK Ltd

Miscellaneous

Agency

Universal Music Operations Ltd *v*. Flairnote Ltd

Subsequent arbitration proceedings

Baldwin Industrial Services *v*. Barr Ltd
Mackley (JT) & Co Ltd *v*. Gosport Marina Ltd

Non-act adjudications

Gennaro Maurizi Picardi t/a Picardi Architects *v*. Cuniberti and Another
Maymac Environmental Services Ltd *v*. Faraday Building Services Ltd
Parsons Plastics (Research & Development) Ltd *v*. Purac Ltd (1) (2)
Project Consultancy Group *v*. Trustees of the Gray Trust

INDEX

adjudication
  commencement of, 15, 16, 18, 75
  conduct of, 57, 60
  contractual provisions, 7, 8
  costs of, *see* adjudicator, costs of; parties'
    costs
  day, 11
  definition of, 3
  non-statutory, 3, 4
  notice of, 41, 42, 76, 77
  objection to, 77–79
  procedures, 34
  refusal to participate in, 32, 33, 60, 61
  right to refer to, 17
  suspension of, 106
  timetable of, 18, 19
adjudicator, 10,
  agreeing an, 26, 27, 83, 84
  appointment of, 43–48
    revocation of appointment, 97
  concerns regarding, 80–82, 98–101, 113,
    114
  costs of, 36–38, 66
    excessive, 66, 112, 113
    payment of, 90
    recovery of, 89,
  directions of, 103, 104
  expert advice to, *see* expert,
    adjudicator's
  fees and expenses of, *see* costs of
  immunity of, 17
  impartiality of, 99, 100
  initiative by, 100
  jurisdiction of, *see* jurisdiction
  powers of, 57–59
  qualifications of, 80, 81

  suitability of, 80, 81
  terms and conditions of, 48, 49
adjudicator nominating body, 9, 10,
  45–50
administration order, 33
administrative receivership, 71
advisers, *see* representation
agreement
  adjudication, adjudicator's, 48, 49, 137,
    138, 141, 144, 145, 147, 155
  binding, 18
  evidence of, *see* writing, agreement in
arbitration, referral to, 4, 16, 18, 92, 116,
  117
arrestment, 30
Association of Consultant Architects Form
  of Building Agreement, 3, 4
Association of Consulting Engineers
  Conditions of Engagement, 7

bad faith, 17, 99, 113
bias, *see* impartiality, lack of
British Property Federation Form of
  Building Agreement, 3, 4
burden of proof, 53

conciliation, 15, 18
concurrent proceedings, 30, 70, 71
construction contracts, 24–26
  relevant, 122
construction operations, 26, 27
  exclusions from Act, 27–29
Construction Contracts Exclusion Order
  (England & Wales) 1998, 28
Construction Contracts (Northern Ireland)
  Order 1997, 27

Construction Industry Council Model
	Adjudication Procedure, 7, 35,
	155–157
contract(s)
	adjudicator's, *see* agreement, adjudicator's
	determination of, 30
	existence of, 19
	law of, 29
	location of, 27
	mixed, 27
	multiple, 30, 31
	repudiation of, 30
	service, 5
	supply only, 5
	terms of, 19
correspondence, 40, 41
counterclaim, 62, 70, 71, 88
court, intervention of, 80, 106

decision, 11
	binding on parties, 3, 108, 109
	challenges to, *see* enforcement of
		decision
	correction of, *see* slips and errors
	enforcement, *see* enforcement of
		decision
	failure to take up, 90, 91
	format of, 62, 63
	lateness of, 107, 108
	obtaining, 64, 65, 90
defence, *see* response to referral notice
dispute(s)
	existence of, 19, 20
	indirect, 21
	meaning of under ICE Conditions of
		Contract, 137
	under different contracts, 31
	under same contract, 30
documentation, 54, 55

enforcement of decision, 4, 66–69
	challenges to, 91–94, 113–116
	statutory demand, 68
	summary judgment, 4, 67
Engineering Construction Contract, 3, 4, 7,
	21, 33, 143–147

engineer's decision, 35
evidence,
	new, 104, 105
	oral, 101, 102
	supporting, *see* documentation
expert
	adjudicator's, 106, 107
	evidence, 55

facsimile, 40
final and binding, 3
finance agreements, 29
financial circumstances, 33, 69, 82, 83, 91,
	92

GC Works General Conditions, 7,
	149–151

hearing, 101, 102
Housing Grants Construction and
	Regeneration Act, 5, 17–29, 57, 58
	operative date of, 21, 22
	possible amendment to, 8, 9
Human Rights Act, 114

impartiality, lack of, 81, 82, 114
inhibition, 68
Institution of Chemical Engineers
	Adjudication Rules, 7, 147–149
	Form of Contract, 7, 147
Institution of Civil Engineers
	Adjudication Procedure, 138–141
	Conditions of Contract, 7, 33, 136–138
insolvency, 69, 71, 93
Insolvency Act, 1986, 33, 69, 71
interest, power to award, *see individual*
	procedures in Part 6
issues, 42
	different, 84, 85
	failure to deal with, 65, 66

joint and several liability, 113
Joint Contracts Tribunal Standard Form of
	Building Contract, 7, 141–143
judicial review, 68
jurisdiction, 11, 65, 78–80, 92, 114, 115

power to determine, 11, *see also individual procedures in Part 6*

labour only contracts, 24
Latham, Sir Michael, 5, 121
litigation, referral to, 15, 18, 116, 117

mandatory pre-step, 3, 4
matter of dissatisfaction, 33, 35, 137, 144
mediation, 15, 18
meeting, 87, 88

natural justice, rules of, 100
negotiation, 15
New Engineering Contract, *see* Engineering Construction Contract
new material, *see* evidence, new
nomination, 43–48, 77
non-compliance, 67, 121, 132
novation agreement, 22

Official Referees Solicitors Association, 151
oral evidence, 101, 102

parties
    designation, 9
parties' costs, 38–40, 89
part of the land, 25
payment
    due date for, 122, 123
    final date for, 124
    notice of amount of, 124
    notice of intention to withhold, 124, 125
    pay when paid clauses, 126, 127
    periodic, 122
    provisions of Act and Scheme, 122–127
    requirements, 6
    right to stage payments, 122
primary activity, 26, 27
Private Finance Initiative, 29
procedural fairness, 100
procedural rules, 34, 58, *see also individual procedures in Part 6*
process, adjudication, 57–60

professional negligence, 47

quantum, 53

reasons, reasoned decision, 63
    requesting, 108, 109, 110
redress sought, 53, 54
referral, referral notice, 9, 50–56
    defence to (response), *see* response to referral notice
    different from notice of adjudication, 84, 85
    duplication of, 31, 32
    lateness of, 56, 57
    withdrawal of, 61, 62
reply, 59, 60
representation, 35, 36, 83
residential occupier/premises, 28
resources, 97, 98
responding party, 9
response to referral notice, 59, 85–87
Royal Institute of British Architects, Conditions of Engagement, 7
Royal Institution of Chartered Surveyors, Conditions of Engagement, 7

Scheme for Construction Contracts
    England and Wales, 6, 7, 35, 49, 51, 58, 59, 132–136
    Northern Ireland, 7, 136
    possible amendment to, 7, 8
    Scotland, 7, 136
set-off, 70, 71
settlement, 18, 32, 89, 116
site inspection, 102
slips and errors, 110, 111
subcontract, forms of, 157
subcontractor, 35, 39, 126–7, 141
submissions, further, 60
summary judgement, 4
suspend work, right to, 125, 126

Technology and Construction Solicitors Association Adjudication Rules, 7, 151–155

value added tax, 63

withholding notice, *see* payment, notice of
    intention to withhold

indiscriminate, 5
writing, agreement in, 22–24
   see also agreement, evidence of